www.wadsworth.com

www.wadsworth.com is the World Wide Web site for
Wadsworth and is your direct source to dozens of
online resources.

At *www.wadsworth.com* you can find out about
supplements, demonstration software, and student
resources. You can also send email to many of our
authors and preview new publications and exciting
new technologies.

www.wadsworth.com
Changing the way the world learns®

The Counselor Intern's Handbook

THIRD EDITION

CHRISTOPHER FAIVER
John Carroll University

SHERI EISENGART
Case Western Reserve University

RONALD COLONNA
Lutheran Children's Aid & Family Services,
Cleveland, Ohio

THOMSON
™
BROOKS/COLE

Australia • Canada • Mexico • Singapore • Spain • United Kingdom • United States

THOMSON
™
BROOKS/COLE

Executive Editor: *Lisa Gebo*
Technology Project Manager:
 Barry Connolly
Editorial Assistant: *Sheila Walsh*
Marketing Manager: *Caroline Concilla*
Marketing Assistant: *Mary Ho*
Advertising Project Manager: *Tami Strang*
Project Managers, Editorial Production:
 Cheri Palmer and Stephanie Zunich

Print/Media Buyer: *Doreen Suruki*
Permissions Editor: *Kiely Sexton*
Production Service: *Anne Draus, Scratchgravel*
 Publishing Services
Copy Editor: *Marilyn Fahey*
Cover Designer: *Bill Stanton*
Compositor: *Scratchgravel Publishing Services*
Printer: *Webcom*

Printed in Canada

1 2 3 4 5 6 7 07 06 05 04 03

For more information about our products,
contact us at:

**Thomson Learning Academic
Resource Center
1-800-423-0563**

For permission to use material from
this text, contact us by:

Phone: 1-800-730-2214
Fax: 1-800-730-2215
Web: http://www.thomsonrights.com

Library of Congress Control Number:
2003103017

ISBN 0-534-52835-X

Brooks/Cole—Thomson Learning
10 Davis Drive
Belmont, CA 94002-3098
USA

Asia
Thomson Learning
5 Shenton Way #01-01
UIC Building
Singapore 068808

Australia/New Zealand
Thomson Learning
102 Dodds Street
Southbank, Victoria 3006
Australia

Canada
Nelson
1120 Birchmount Road
Toronto, Ontario M1K 5G4
Canada

Europe/Middle East/Africa
Thomson Learning
High Holborn House
50/51 Bedford Row
London WC1R 4LR
United Kingdom

Latin America
Thomson Learning
Seneca, 53
Colonia Polanco
11560 Mexico D.F.
Mexico

Spain/Portugal
Paraninfo
Calle/Magallanes, 25
28015 Madrid, Spain

We dedicate this book to our families, friends, and students.

C.F.:   To Hubert Jones, Jake Jones, Dan Noday,
        Marc Noday, and Cory "C. J." Jones

S.E.:   To Seth, Laurie, Julie, Jonathan, and Jodi

R.C.:   To Cathy, Andrew, Julie, and Stephanie Colonna

Contents

Preface

This basic guide is designed to assist counseling students and others in the helping professions through the entire experience of internship, often the last requirement of a degree program. We suggest its purchase before the internship placement because Chapter 1 gives suggestions for selecting a site. Furthermore, many will find it helpful as a study guide for licensure and certification examinations, and seasoned professionals may use it as a quick reference. We include overviews of basic treatment modalities, psychological testing, and psychopharmacology, as well as chapters on the clinical interview and ethical considerations—subjects that are particularly relevant to interns. Our fields are counseling and psychology, yet students in related areas of the behavioral sciences may find the book helpful.

Among us we have supervised field placement students in a variety of settings for more than 50 years, so we are able to maintain a practical perspective in this book. Our varied backgrounds include teaching, administration, research, and clinical work. We encourage students to review what they have learned while they progress to new material. We explain the components of internship in detail, and we urge students to use these components with enthusiasm as well as caution. It is important to leave no stone unturned during the internship process, but we provide only the stones, respecting the student's responsibility for the turning.

Key Changes in This Third Edition

- All chapters are updated with the latest information based on current research and theory.

- Chapters are reordered for increased user friendliness—for example, with the understanding that counselor self-assessment is indicated earlier in the experiential process whereas competencies occur later.

- Diversity and multicultural considerations are incorporated throughout the text.

- New appendices include a student log/journal, technological competencies, and a comprehensive table of psychotropic medications.

- Added competencies consist of technology, managed care, and diversity issues.

- The extensively revised psychopharmacology chapter adds sections on children's, herbal, Alzheimer's, and sleep medications, and on brain structure and function, and the chapter now includes helpful tables and Web sites for further information.

- The testing chapter lists sections on instruments for use with children and clinical populations.

- The chapter on ethics addresses Internet counseling issues.

- The chapter entitled "Understanding How to Help" has been updated, based on the latest treatment-outcome research findings, and includes an entirely new section on postmodern therapies with discussions of narrative therapy, solution-focused brief therapy, and dialectical behavior therapy.

- Sample self-assessment questions can be downloaded from the Wadsworth Counseling Web site at http://wadsworth.com/counseling_d/. Click on Student Book Companion Sites, and then click on the icon for this book.

Acknowledgments

Some of the information we offer emerged from institutions involved in providing field experiences to interns; in particular, we thank the John Carroll University Community Counseling Program Committee. A number of students and colleagues contributed valuable assistance while we were writing this book. They include Paula Britton, Ph.D.; Rhonda Harrison; Dawn Harsch, M.S.; David Helsel, Ph.D.; Keith Mesmer, M.Ed.; Allison Metz; Eugene O'Brien, Ph.D.; Cynthia Osborn, Ph.D.; John Ropar, Ph.D.; Debbie Vernon, M.Ed.; and Michael Yaksich. We are grateful to Brooke M. Wolf, M.D., clinical instructor in psychiatry at Case Western Reserve University School of Medicine, for her expert advice concerning our chapter on psychotropic medications. We also wish to acknowledge Gerald Corey, Ed.D., for his inspiring writing.

We also appreciate the efforts of the following reviewers: Fred T. Ponder, Texas A&M University–Kingsville; Robert J. Kennerley, University of Central Florida; Norman Cohen, Bowie State University; and Michael D. Loos, University of Wyoming..

The staff at Brooks/Cole–Wadsworth was enormously helpful. We appreciate the encouragement and support of Lisa Gebo, Julie Martinez, Sheila Walsh,

Stephanie Zunich, Cheri Palmer, Anne Draus, and all of the other hard workers who contributed to this edition.

　　We look forward to your comments. Please write to us at Brooks/Cole–
Wadsworth Publishing Company, 511 Forest Lodge Road, Pacific Grove,
CA 93950.

Christopher Faiver
Sheri Eisengart
Ronald Colonna

The Counselor Intern's Handbook

1

Getting Started

Your internship is generally the culmination of the academic sequence leading to your degree in counseling. It is an exciting, challenging time, and many students anticipate it with mixed emotions. Internship students often find themselves facing unfamiliar situations, engaging in intense encounters, and processing powerful feelings, which all lead to increased introspection, personal reassessment, change, and growth. The internship may be your first opportunity to interact independently with real clients in a professional capacity as a counselor, as well as your first experience in scrutinizing those interactions during a formal supervisory hour. This book is a step-by-step guide that will remove some of the anxiety from your internship and allow you to relax and enjoy the many rewarding, wonderful moments you are about to experience as a counselor intern.

Ideally, your internship should give you a supportive, structured learning environment for acquiring clinical experience and practical on-the-job training. Unfortunately, the reality can differ. You will be called on to synthesize material from previous coursework, to utilize theories and techniques, and to begin to develop a personal and professional style of relating effectively to clients, clients' families, agency staff members, and other mental health professionals.

During your counseling internship, you will work under the direct supervision of a licensed counselor, social worker, psychologist, or psychiatrist. You will meet for regular supervisory sessions to review your internship experience, as specified by your state licensure/certification board. Ideally, your field placement supervisor should be an experienced clinician who will provide a flexible learning experience tailored to meet your individual needs by encouraging you and continually assessing your strengths and weaknesses as well as asking for your feedback. The actual number of hours you spend in direct contact with clients, as well as the total number of supervisory hours, should always fulfill state licensure/certification requirements. For example, the Council for Accreditation of Counseling and Related Educational Programs (CACREP) specifies that 40 percent of the required 600 internship hours should be spent in direct client contact and that the intern must average 1 hour of individual and/or triadic supervision (meaning two interns, one site supervisor) and an average of 1.5 hours of group supervision per week (usually performed by a program faculty member). So a graduate student fulfilling a 600-hour internship would need 240 hours of client contact and 30 hours of supervision to have the internship meet these national standards. Many states have adopted these standards for licensure in counseling.

In addition to on-site experience and supervision, you will most likely meet on campus with other interns to discuss your placement experiences. Most students appreciate this group time with their peers to share their feelings, their frustrations, and their accomplishments, and they find the support this group offers especially helpful during their internships. You may also meet with a supervisor from your counseling program (usually one of the department or program professors) for further processing of your individual experiences. You will typically be required to keep a detailed log or journal of daily on-site activities (see Appendix I) and to prepare two or more formal case studies for class presentation. Additional research projects or reading relevant to counseling may be assigned, as well. The agency and university supervisors will evaluate your performance and potential as a counselor, and they should discuss their evaluations with you both midway through the internship and again at completion of the experience.

SELECTING YOUR INTERNSHIP SITE

Selecting your placement site is the first, and perhaps the most important, step in your internship experience. The selection process integrates four factors:

- Your own interests and needs
- The field placement guidelines of your university counseling or human services program
- The state requirements for on-the-job experience for professional counseling licensure/certification
- The didactic and experiential opportunities afforded at the placement site

These four factors are interdependent, and selecting a placement site is a dynamic process of exploring and matching these various criteria to find a good fit. Some counseling and human services programs have formal placement relationships with preapproved agencies, which may limit selection to a certain extent; however, this may also help provide a satisfactory field experience. Other programs delegate most of the responsibility for securing field placements to the individual student. In either case, you should first try to determine your own interests, needs, and expectations when you begin the process of choosing a placement site. (A few programs assign interns to a specified site with little or no input on the part of the student.)

Your Interests and Needs

You may already have a fairly good idea about what types of people you enjoy working with (for example, children, adolescents, adults, or the elderly) and what kinds of problems you would like to deal with (for example, substance abuse, child and family concerns, career counseling, or mental health issues such as depression and anxiety). Perhaps you prefer group or family work to individual counseling. Moreover, even at this point in your career, you may have an idea of where you'd like to work after graduation—dependent upon the job market, of course. Knowledge about your own preferences gives you some direction when you begin looking for an internship site because you can limit your choices to those places where you are certain that you have a keen interest in agency clients and services. You may also use your internship as an opportunity to try something new, to deepen your self-awareness, and to enlarge your scope of experience. For example, if you have been working with abused children but have not had any helping interactions with chemically dependent adults, your internship may provide a chance to obtain practice in a different area.

Agencies serve populations and provide services that are either homogeneous (such as drug and alcohol treatment centers or child welfare agencies) or heterogeneous (such as community mental health centers or psychiatric hospital units). If you are unsure about your interests, or if you do not prefer working with one type of client or one particular type of problem, then you may want to consider heterogeneous internship sites, which will offer you a wider range of experience. We suggest that you consider the opportunity of working with persons from a wide variety of cultures and ethnic backgrounds.

In beginning the selection process, you should also consider whether an inpatient or an outpatient setting would satisfy more of your needs and interests. The hospital inpatient setting will typically be fast-paced and high-pressured, with a rapid client turnover. If you select a psychiatric inpatient unit for your internship, you will most likely come into contact with a great many clients, most of whom are in acute distress and are manifesting serious psychopathology. Working with the psychiatric inpatient is similar to crisis intervention in that your counseling relationship and interventions help the client stabilize and return to a prior level of functioning. The hospital unit lets you interact with other health and mental health professionals as a member of an interdisciplinary treatment team, as well as learn to relate to clients who have complicated psychiatric disorders, and these experiences are valuable and interesting.

However, the intensity of the emotional upset and the extent of the problematic behaviors encountered in a hospital inpatient unit may not suit all counseling students.

Long-term residential treatment facilities provide clients who generally have chronic, rather than acute, problems or difficult management issues and who require a more structured or supportive environment. One advantage of this type of internship is the extended counseling relationships that you may develop with your clients, who are likely to remain at the facility throughout your placement. In addition, you may have special opportunities to observe, learn about, and interact with your clients as they go about their daily activities. An internship at a long-term treatment center for children or adolescents can be particularly rewarding in these respects.

If you choose an outpatient setting, your internship experience will vary according to the particular agency. Some agencies may let you work as a member of an interdisciplinary team and participate in treatment conferences, whereas at other agencies you may interact only with your supervisor to discuss cases. Most community mental health centers must by law provide services in such areas as intake assessment, emergency care, consultation and education, research and evaluation, individual and group counseling, and after-care planning. The outpatient setting will usually, but not always, serve clients who are less distressed and less acute than those in the inpatient unit. As an intern in an outpatient center, you may have more opportunities to use a variety of counseling techniques and activities because the clients are generally higher functioning and are able to cope more effectively with daily living tasks. However, if you work in the intake or emergency department of a community mental health center, you may deal with highly upset clients or clients in crisis, who may require immediate hospitalization. In addition, many of your clients may have chronic problems or be "repeaters" who require continued support and management.

Another aspect of your internship needs may involve financial considerations. At some agencies interns are paid for their services; however, these sites will be limited, and therefore your choices will be somewhat restricted if you require financial compensation during your internship. Unfortunately, many excellent internships are unpaid positions. Some agencies will hire a graduate counseling student to perform a paid job, such as case manager or mental health technician, and then allow the student to set aside a designated number of unpaid hours specified as internship hours, when the student assumes tasks and responsibilities relevant to the internship. Some students find this arrangement satisfactory, whereas others report that they are frustrated with their limited counseling and supervised time.

You should also consider whether a particular agency's schedules mesh with your own needs. For example, some agencies offer group therapy sessions on several evenings during the week, which may conflict with your personal responsibilities. Attempt to ask questions about this up front when interviewing for the internship placement as well as the kinds of clinical activities and responsibilities expected of interns in that setting.

Your Counseling Program Guidelines

As you begin exploring possible internships, be certain that your potential site fulfills all the requirements of your counseling program and state licensing board. Most programs print clear guidelines listing the expectations and regulations for student internships. For example, your counseling program guidelines may specify that interns acquire experience in treatment planning, in individual and group counseling, in case management, and in discharge planning, as well as gain an understanding of agency administrative procedures. You should be sure that you will be able to satisfy all the necessary program requirements at each internship site. In addition, you must determine whether you are covered by malpractice insurance provided by your program. Your counseling program guidelines should specify whether students are covered. If you need to purchase liability insurance, professional organizations such as the American Counseling Association (ACA) provide coverage at reduced rates for students. It's always a good idea to have liability insurance.

State Licensure/Certification Requirements

Many state counselor licensure/certification boards allow credit for supervised hours accrued during your internship, even before you complete your graduate program in counseling. However, specific rules and regulations address, for example, whether your experience may be paid or unpaid, the hours and nature of supervision, the relationship between supervisor and supervisee, and the intern's scope of practice. In addition, state boards have formal procedures and policies for registering supervised counseling experience.

As soon as you begin selecting an internship site, you should write or call your state counseling regulatory board to obtain all documents, applications, and instructions pertaining to counselor licensure and certification. Then read through everything carefully and consider whether you will be able to fulfill the state requirements for supervised counseling experience at each internship site. In addition, once you have selected a site, be sure to follow through on all formal procedures to register your hours with your state board so that you can get credit toward state licensure or certification for your internship. Make certain that your prospective supervisor meets requirements for supervision by contacting your state board. In the cases of the few states with no licensure or certification, we recommend following the requirements of the National Board for Certified Counselors (2002). Appendix G provides contact information for state and national credentialing boards.

Experiential and Didactic Opportunities of Each Site

Your internship is a critical part of preparing for your career as a professional counselor, and you should carefully examine and analyze what kinds of educational opportunities you will be likely to experience at each potential site. Will you have direct client contact, such as doing intake assessments and conducting individual or group counseling, or will much of your time be spent running errands, such as filing papers and operating the copying machine? Will you be treated as a colleague and a valued member of the treatment team, or will the other professionals discount your input because you are a student? Will you have the support you need, and will other staff members be willing to answer questions and offer help if you ask? Will you be invited to attend staff meetings and in-service educational sessions? Your prospective supervisor will be responsible, in large part, for delineating your internship activities. Therefore, you should thoughtfully consider the personal and professional qualities you hope to find in this individual. Your supervisor also must have the time, the interest, and the commitment to teaching interns, plus an understanding of the special nature of the supervisory relationship.

SOURCES OF INFORMATION

As you begin the process of selecting your internship, you will need to assemble a list of potential sites. You may already have some thoughts about where you would like to do your internship. However, if you have no idea where to begin, we suggest a preliminary discussion with your counseling program coordinator of field placements. The coordinator should be able to provide helpful suggestions and insights concerning prospective placement sites, as well as names of alumni of your program who are currently working at various sites who would be good people to contact for further information. It may also be a good idea to sit down with your faculty adviser, who ideally knows something about your personality and needs, and who may be able to recommend several possible placements. Many universities have a list of approved internship sites.

Networking with other students who are currently involved in a field placement or who have recently completed one is an invaluable way to gather information about potential sites. In this way you may hear of several interesting sites that you did not know were available. You will probably also learn a great deal about a particular site's activities, schedule, staff, and atmosphere by talking to students who have worked there. Some of these same questions may be answered by site supervisors at the informational interview, which is discussed in the next section; however, it is often helpful to have another student intern's perspective as well.

One additional source for potential field placement sites is the classified ad section of your local newspaper. The help wanted advertisements, under such headings as "social services," "medical settings," "counselors," "education," and "mental health," often provide possible leads. You can telephone each site and ask whether an internship or field placement program is available and whom to contact for further information. In addition, many funding agencies, such as the United Way or local mental health board, publish listings of their affiliated agencies. Often these listings, which usually include agency service descriptions as well as contact persons, provide good leads for internships. Web sites may also be of use.

INFORMATIONAL INTERVIEWING

You may acquire awareness of your own needs, interests, and expectations through course readings and experiences, as well as through actual visits to potential placement sites. Discussions with supervisors and clinicians at the site will enable you to clarify the four factors to consider in choosing your internship: (1) your own needs and interests, (2) your counseling program guidelines, (3) state regulations concerning counseling licensure/certification, and (4) the educational opportunities at each site. A visit to a potential site helps you determine whether that particular internship site is a good fit and meets most of your expectations. This process, known as informational interviewing, can give you invaluable firsthand exposure to each site's unique environment.

To prepare for an informational interview, we suggest that you compose or update a resume outlining your educational and professional experience, including volunteer activities relevant to counseling or human services. (A sample resume is provided in Appendix A.) Take along a copy of your resume, plus a portfolio containing samples of written work (such as case studies or research projects from courses), and also a copy of your counseling program's field experience regulations, including agency guidelines, course requirements for students, and evaluation procedures.

You should be prepared to answer, as well as to ask, questions when you go on your informational interviews; the professional at the placement site may look on this interview as an ideal opportunity to learn something about your personality and ability. Future supervisors have asked potential student interns a wide variety of questions, ranging from "What is your theoretical orientation?" to "What is your favorite restaurant?" to "Have you ever been in therapy yourself?" Your attitude toward being questioned and your style of relating to the interviewer may be more important than the actual answers to those questions. You may want to ask about agency philosophy, goals, services, and history as well as such practicalities as hours, types of supervision, and so forth.

Following the interview, we suggest that you send a brief thank-you note indicating your interest, if any, in the placement and noting that you will follow up with a phone call in one or two weeks to explore the possibility of arranging an internship at that site. (Appendix B provides a sample thank-you note.)

PLACEMENT DESCRIPTIONS

Once you have listed potential placement sites and have visited the most promising choices, you can compare aspects of each placement to find one that best meets most of your needs and interests, fulfills your academic program specifications, satisfies as many of the state licensing/certification requirements as possible, and provides optimal educational opportunities. To facilitate selection, you may find it helpful to prioritize criteria. For example, you may decide that it is more important to you to work with one specific client population than it is for you to work as an interdisciplinary team member. In addition, keep in mind that the goal is to find a good fit, rather than the perfect choice, so try to be flexible in assessing each placement site. Remember that no site is ever perfect.

Placement descriptions provide key information for you to consider, in conjunction with other data, because they represent the supervisor's or administrator's standards and expectations for each placement experience. You may be able to pick up printed placement descriptions in the program office at your college or university, in the classified ad section of the newspaper, or in various agency offices. You may also find it helpful to write notes on the verbal job descriptions given at informational interviews or during discussions with other students, advisers, or professors. These notes can be reviewed and compared later on. (A sample internship announcement is provided in Appendix C.)

As you review these placement descriptions to select your internship site, you may find it helpful to consider the following questions:

- What kinds of people do I enjoy working with? Will I be likely to work with this population at this placement site?

- What types of problems might I want to deal with? Will I encounter these types of problems here?

- Is the experience paid or unpaid?

- Does the placement allow for credit toward state licensure or certification requirements? Some state licensing/certification boards, for example, specify that the student must receive at least 1 hour of supervision for every 20 hours of client contact. You may want to call the state credentialing board to check on the status of both the site and supervisors.

- How much will I actually be working with clients, rather than observing or running errands?

- Will I receive adequate individual supervision? Is the supervisor credentialed at the independent practice level of licensure or certification? Having met the supervisor, do I feel that we will be able to maintain a good working relationship?

- Am I covered by liability insurance? Most institutions of higher education or agencies themselves carry policies that insure students during their internships. In cases where the university, college, or agency does not provide insurance, students should seek coverage through other sources, such as professional organizations (for example, the ACA).

- What is the general atmosphere at this placement site? Is it formal or informal? Will I be comfortable working here? Will I be welcomed as a colleague?

- What other mental health professionals are on staff? Will I be directly involved with psychiatrists, psychologists, social workers, counselors, nurses, teachers, or child care workers? Will I be part of a treatment team?

- Do I have a backup plan if this placement falls through? We suggest interviewing in more than one place.

GETTING READY FOR THE PLACEMENT

Finally! You think you have found an internship site that is a good fit for you, and your future supervisor has told you that you have the position. You wonder what to do next. First, you will need a written agreement for the placement, signed by your future supervisor, specifying the dates of your internship. Next, you should follow the prescribed state licensing/certification procedures to obtain credit for your supervised hours. This process usually involves both you and your supervisor completing paperwork describing the proposed scope of practice, the setting, the number of hours of work and of direct supervision, and the supervisor's areas of expertise. Keep copies of all documents.

Your meeting with your supervisor, when these official forms are completed and signed, is a good time to ask, "What can I do, before I actually begin work, to best prepare myself for this placement?" Your supervisor may be able to offer some advice, for example, by suggesting a few relevant books for you to read or by noting that you should thoroughly review the *Diagnostic and Statistical Manual of Mental Disorders* before beginning the placement experience. At this meeting with your supervisor, you may also want to discuss the actual scheduling of your work hours so that an agreement acceptable to both of you can be worked out. Some students have found it useful to discuss dress codes, if applicable, and whether they will need to bring any special equipment or supplies. For example, one inpatient unit we know requires all counselors to use pink highlighters on their patient chart notes to make these notes easy to see. At this internship site, interns come prepared with several pink highlighters each day.

You may also find it helpful, during the time before you begin your internship, to review academic coursework, concentrating on those areas that might require extra attention. The National Board for Certified Counselors (NBCC) provides coursework descriptions in the 10 areas of study relevant to counseling, and these may be used as guidelines for self-assessment (see Appendix D for the NBCC course descriptions). We also suggest a thorough reading of the American Counseling Association *Code of Ethics and Standards of Practice* (1995) so that you will be ready to act in a professional manner in all situations.

OTHER PLACEMENT VENUES

Although the primary focus of this book is the mental health counseling realm, other related settings offer unique and valuable experiences for the counseling intern. These include school counseling, higher education counseling, and employee assistance counseling settings.

The school counselor intern, by virtue of his or her placement setting, is on the front lines of the counseling profession. He or she often confronts issues at the grassroots level, with home problems frequently becoming school problems. School counselors sometimes face parents' desperate, but unfair, demands to do intensive therapy or to handle children with severe psychological problems.

Moreover, unfortunately, it has been our experience that some school counselors are asked to perform duties and assume roles that appear to directly conflict with counseling duties and roles. For instance, some school counselors are asked to become disciplinarian, assistant principal, test administrator (as sole function), class scheduler, hall monitor, or (even) baby-sitter. Very often these tasks incorporate an authoritarian or clerical role, which conflicts with training as a nonjudgmental and skilled professional counselor. Role conflict may result, causing consternation, role confusion, and stress.

Ideally, the school counselor does just what the title denotes: he or she counsels. Implicit in the role of a school counselor are various functions appropriate to his or her professional training, such as career, vocational, and college guidance and testing; counseling regarding situational concerns; referral to other professionals when appropriate (such as to the school psychologist or other mental health professionals external to the school for problems that appear to be beyond the scope of the school counselor's role and training); parent conferences; teacher consultation and training in counseling areas; and so on. Serious psychological and psychiatric disorders should be treated in a clinical setting.

Higher education counseling can also be a rewarding experience for the counselor intern. Many placements are possible at the college or university level, including the counseling center, career and placement office, student development center, office of multicultural affairs, and other student-related departments such as housing and athletics. Because of the diversity of placement opportunities, work can vary according to the particular setting—from clinical mental health counseling at the counseling center to career and vocational counseling at the career services or placement center to performance enhancement in the athletic department.

Employee alcoholism programs of the 1940s and 1950s have been organized into employee assistance programs (EAPs) in more recent decades as part of a "benefit package" to employees. Employee assistance counselors assist in identifying and resolving productivity problems associated with employees concerned with personal troubles such as health problems, marital or family issues, financial concerns, alcohol or drug problems, legal or emotional problems, and stress. Often employee assistance counselors provide consultation and training in job performance areas, referrals for diagnosis and treatment, crisis intervention, linkages between workplace and community resources, and follow-up services.

Whereas the early development of EAPs involved organized labor, management, and Alcoholics Anonymous, in more recent times EAPs have become more professional, with many counselors, social workers, and psychologists becoming certified in employee assistance. In addition, many states are moving to license individual EAP providers.

A CHECKLIST FOR GETTING STARTED

To help you complete all the steps in selecting your internship site, we have compiled the following checklist:

1. Determine your own interests, needs, and expectations for the internship.

2. Review your counseling or human services program regulations for internships.

3. Ascertain any state requirements for supervised work experiences related to counseling licensure/certification.

4. Think about what kinds of educational opportunities you hope to experience during your internship, and what qualities you hope to find in your supervisor.

5. Make an appointment with your faculty adviser to discuss placement possibilities.

6. Make an appointment with your program placement coordinator to acquire further information.

7. Locate additional sources for information about potential internship sites, including other students, former interns, newspaper advertisements, and funding agencies.

8. Compose or update a resume and make several typed copies.

9. Arrange informational interviews as often as possible.

10. Follow up on each interview with a thank-you note and, a week or two later, a phone call.

11. Obtain and compare job descriptions for the internship sites that interest you.

12. Try to find a good fit for the placement, keeping in mind your own needs and interests, your program requirements, state licensing/certification regulations, and the unique experiences offered by and the individual characteristics of each internship site.

13. Secure a final agreement with the supervisor after you have selected a placement site.

14. Complete all official forms to register your supervised experience with the state licensing/certification board.

15. Determine the necessity of obtaining liability insurance. The American Counseling Association offers student members coverage at minimal cost (800-347-6647).

16. Ask your future supervisor how you can best prepare for your placement. Follow through on his or her suggestions.

17. Assess your academic record for areas relevant to the internship that you need to review.

18. Set up a schedule to read and review material so that you will begin your placement feeling as competent and comfortable as possible.

REFERENCES

AMERICAN COUNSELING ASSO-
CIATION. (1995). *Code of Ethics and
Standards of Practice.* Alexandria, VA:
Author.

NATIONAL BOARD FOR CERTI-
FIED COUNSELORS. (2002). *NCC
Application.* Greensboro, NC: Author.

BIBLIOGRAPHY

BRADLEY, F. (Ed.). (1991). *Credentialing in
counseling.* Alexandria, VA: American
Counseling Association.

COLLISON, B., & GARFIELD, N.
(1990). *Careers in counseling and development.* Alexandria, VA: American Counseling Association.

COREY, G. (2001). *Manual for theory and
practice of counseling and psychotherapy*
(6th ed.). Pacific Grove, CA: Brooks/
Cole.

COREY, G. (2001). *Theory and practice of
counseling and psychotherapy* (6th ed.).
Pacific Grove, CA: Brooks/Cole.

EMPLOYEE ASSISTANCE PROFES-
SIONALS ASSOCIATION. (1992).
Standards for employee assistance programs.
Arlington, VA: Author.

HEPPNER, P. (Ed.). (1990). *Pioneers in
counseling and development: Personal and
professional perspectives.* Alexandria, VA:
American Counseling Association.

HERR, E. (1999). *Counseling in a dynamic
society: Opportunities and challenges.*
Alexandria, VA: American Counseling
Association.

HERRING, R. (1998). *Career counseling in
schools: Multicultural and developmental
perspectives.* Alexandria, VA: American
Counseling Association.

KOTTLER, J. (1997). *Finding your way as a
counselor.* Alexandria, VA: American
Counseling Association.

MYERS, W. (1992). *Shrink dreams.* New
York: Simon and Schuster.

PECK, M. (1999). *The road less traveled.*
New York: Simon and Schuster.

PEDERSEN, P., & CAREY, J. (1994).
*Multicultural counseling in schools: A
practical handbook.* Boston: Allyn &
Bacon.

SCHMIDT, J. (1998). *Counseling in schools:
Essential services and comprehensive
programs* (3rd ed.). Boston: Allyn &
Bacon.

SIEGEL, S., & LOWE, E. (1999). *The
patient who cured his therapist: And other
stories of unconventional therapy* (2nd
ed.). New York: Dutton.

SKOVHOLT, T., & RONNESTAD, N.
(1992). Themes in therapist and counselor development. *Journal of Counseling and Development, 70*(4), 505–515.

SUSSMAN, M. (1992). *A curious calling:
Unconscious motivation for practicing
psychotherapy.* Northvale, NJ: Jason
Aronson.

TALLEY, J., & ROCKWELL, W. (1986).
*Counseling and psychotherapy with college
students.* Westport, CT: Praeger.

WALLACE, S., & LEWIS, M. (1990).
Becoming a professional counselor. London: Sage.

WOODARD, D., COVE, P., and KOMES,
S. (2000). *Leadership and management
issues for a new century: New directions for
student services #92.* New York: Jossey-
Bass.

YALOM, I. (1974). *Every day gets a little
closer.* New York: Basic Books.

YALOM, I. (1989). *Love's executioner.* New
York: Basic Books.

2

Along the Way

A Counselor Self-Assessment

We are in the business of introspection, reflection, analysis, synthesis, and personal action. We personify what Freud (1943) called the *talking cure:* our words are our product. It certainly behooves us to look at ourselves as we move through the process of creative self-examination with our clients, not only during this internship, but throughout our careers. Although this self-examination can take considerable courage, it can also help us appreciate how our clients may feel as they expose themselves to us in the therapeutic milieu. We have noticed that, as counselors, we sometimes become desensitized and myopic to the justifiable defenses evident as a client self-discloses issues and concerns, often at our prompting. Our own thorough self-examination may foster increased capacity for empathy and personal and professional refinement. Our own issues and countertransferences may surface in this process, in addition to other possible character quirks (perhaps even our own neuroses). All personal issues are worthy of inspection. For this reason, we recommend personal therapy for all counselors as an adjunct to training. It helps to see the process from the other side.

Bernard and Goodyear (1998) advocate that supervisors teach interns the principles of both peer and self-evaluation and expect such an examination before each supervisory session. Corey and Corey (1998) likewise advise that counselors-in-training perform self-exploration so as to better understand their clients' issues. Moreover, they note that for many students, self-exploration is a more difficult task than that of apprehending their clients' situations. Benjamin (1987), who himself conquered a major obstacle—his blindness—shares his personal philosophy of self-knowledge, which involves trust in our own feel-ings, ideas, and intuitions. Further, as we become at ease with personal self-exploration, we not only nurture our own personal growth but also enhance our understanding of clients.

In this chapter we offer the beginnings of the self-examination process for counselors-in-training. Our purpose is to suggest areas of personal reflection and inner dialogue. Certainly, who we are as people influences our choice of profession and how we conduct ourselves as professionals. Thus, we first ask you to look at yourself from a personal vantage point, examining your values, beliefs, needs, priorities, characteristics, and biases. Next, assess your professional role, your career goals, and your personal traits that have influenced your career choice. We realize that it's not possible to divorce the personal from the profes-sional; in fact, if we are fortunate, our career is an important extension of our personality. However, for the sake of clarity, we posit a somewhat artificial di-vision. The chapter presents questions in an attempt to elicit some inner prob-ing. We do, however, advise you to seek out trusted fellow students or profes-sors for pertinent discussion, feedback, and validation.

WHO AM I AS A PERSON?

1. How do I assess my developmental history up to this point of my life? What were the high and low points?
2. When did I realize I was an adult? How did I handle this realization?
3. What are my five best qualities?
4. What five areas of my life do I need to improve?
5. If an interviewer asked me what my basic philosophy of life is, what would I answer?
6. Is my glass of water half full or half empty? Why?
7. What pervasive mood do I find myself in most of the time?
8. What do I think about people in general?
9. What role do my religion, culture, ethnic values, gender, and sexual ori-entation play in my view of life?
10. How do I respond to those who are different from me? How open am I to multiculturality?
11. On the Meyers Briggs Type Inventory, what type am I?
12. Who are my heroes?
13. Who is the most creative person I know? Why? How can I nurture cre-ativity in my life?
14. What is maturity?
15. What are my personal goals and objectives?
16. How do I cope with life's challenges and personal stress?
17. Am I past-, present-, or future-oriented?

18. Where do I lie on the continuum of cooperation versus competition?
19. Who and what influenced my life?
20. What would my best friend say about me?
21. What is the biggest criticism people have of me?
22. What three adjectives best describe me?
23. Have I examined what Jung (1980) calls my "shadow"? What do I see in it?
24. Am I open to my own potential needs for counseling?
25. How do I ground and center myself? What do I do for my self-care?
26. Do I rely on my intuition? Why or why not? How do I rely on it?
27. Do I take time to play? How do I play?
28. What are my spiritual beliefs? How do they influence me as a person?
29. Do I possess the personal attributes to be an effective counselor?
30. How do I balance meeting my own needs and meeting the needs of others?

WHO AM I AS A PROFESSIONAL?

1. What are my reasons for becoming a counselor?
2. Do I feel my emotional issues will be addressed and resolved by becoming a counselor?
3. What is my need to be a counselor?
4. What makes me think that I will be an effective counselor?
5. What are my countertransference issues? How do I handle my own issues when they emerge?
6. What do I expect from clients?
7. What do I expect from my profession?
8. What do I anticipate getting from colleagues?
9. What are my professional strengths and weaknesses?
10. What are my professional goals and objectives?
11. What would my fellow students say about me?
12. What would my professors and supervisors say about me?
13. With what type of clients do I wish to work? Why?
14. With whom will I not work? With what issues? Why?
15. Do I know when and where to refer?
16. How would I handle stress and burnout?
17. How do I handle praise and criticism of my work?
18. Do I fully attend to my clients? How so?

Every time you interact with your clients, your professional and personal attributes and self become evident. As you build rapport with your clients, they come to value the way you present yourself. Clients want the assurance of your commitment to help them. The self-assessment process empowers you: The more you know about yourself, the more control you have over your life. This control can be passed along to clients. Refinements come about through continued interaction with colleagues, clients, and others with whom we have daily contact. Sample self-assessment questions can be downloaded from the Wadsworth Counseling Web site at http://wadsworth.com/counseling_d/. Click on Student Book Companion Sites, and then click on the icon for this book. You may wish to download the self-assessment form and use it as a periodic check throughout your internship experience. If your answers change over time, consider the implications these changes represent

Be careful not to expect too much of yourself too early in your career. For this reason, we suggest being gentle with yourself; let the natural process of your professional development follow its course. Be yourself. Do not attempt to play a role or adopt what Jung (1980) called a persona. Confidence should build over time. Your self-assessment enhances the formal skills, techniques, and theories you have learned in this evolving process. Finally, knowing what you can and cannot do for your clients helps you grow and strengthens your knowledge base. The ACA *Code of Ethics and Standards of Practice* (1995) encourages you to transfer clients to another professional if you determine that you cannot assist your clients. You cannot be everything to everybody!

Being guide and mentor—and perceived rescuer—is not an easy task. The journey of life is not an easy one, nor is it fair. Our clients know this, although some clients expect that life "should" be fair. At times it may be difficult for us as well as our clients to maintain a positive attitude toward humanity and even existence itself. Yet we must not view life's journey as tedious, threatening, and dismal. We recall the words of a former colleague: "All we do in this lifetime is rearrange deck chairs on the *Titanic*." Is there more than this? We believe so. We have, however, no empirical proof, only a conviction that humanity is basically good and that we have some freedom to intervene in the condition of humankind, albeit limited. Maintaining a personal attitude of hope fosters hope in our clients; exhibiting a basic trust in humankind can engender a sense of inward and profound trust in our clients; maintaining expectations for improvement serves as impetus for change; and demonstrating positive regard for those with whom we come in contact affirms and sustains them in their humanity and dignity. It is these beliefs for which we stand.

REFERENCES

BENJAMIN, A. (1987). *The helping interview with case illustrations.* Boston: Houghton Mifflin.

BERNARD, J., & GOODYEAR, R. (1998). *Fundamentals of clinical supervision* (2nd ed.). Boston: Allyn & Bacon.

COREY, M., & COREY, J. (1998). *Becoming a helper* (3rd ed.). Pacific Grove, CA: Brooks/Cole.

FREUD, S. (1943). *A general introduction to psychoanalysis*. Garden City, NY: Garden City Publishing.

JUNG, C. (1980). *The archetypes and the collective unconscious* (R. Hull, Trans.). Princeton, NJ: Princeton University Press. (Original work published 1902.)

BIBLIOGRAPHY

BRADSHAW, J. (1992). *Homecoming: Reclaiming and championing your inner child*. New York: Bantam.

CAMERON, J. (1992). *The artist's way: A spiritual path to higher creativity*. New York: Putnam.

CLINESS, D. (1996). *The journey of life*. Poland, OH: Banausic.

COREY, G., & COREY, M. (1997). *I never knew I had a choice: Explorations in personal growth* (7th ed.). Pacific Grove, CA: Thompson Learning.

DAY, L. (1996). *Practical intuition: How to harness the power of your instinct and make it work for you*. New York: Villard.

EMERY, M. (1994). *Intuition workbook*. Englewood Cliffs, NJ: Prentice Hall.

FULGHUM, R. (1993). *All I really need to know I learned in kindergarten: Uncommon thoughts on uncommon things*. Westminster, MD: Faucet.

FULGHUM, R. (1996). *From beginning to end: The rituals of our lives*. Columbine, NY: Faucet.

MOORE, T. (1994). *Care of the soul: A guide to cultivating depth and sacredness in everyday life*. New York: HarperPerennial.

MOORE, T. (1994). *Meditations: On the monk who dwells in daily life*. New York: Harper Collins.

MOORE, T. (1994). *Soul mates: Honoring the mysteries of love and relationship*. New York: HarperPerennial.

PECK, M. (1985). *The road less traveled*. New York: Simon and Schuster.

PECK, M. (1998). *The road less traveled and beyond: Spiritual growth in an age of anxiety*. New York: Touchstone.

3

The Site Supervisor

GENERAL RESPONSIBILITIES

The internship site supervisor plays a highly significant role in, and holds many responsibilities for, your training. With this individual, you will begin to demonstrate the knowledge and skills you have acquired through your formal academic training. The professional and personal relationship you establish with your site supervisor will set the pace, direction, and tone for your internship and, perhaps, for your counseling career. His or her willingness to supervise you becomes a commitment to assist you in attaining and maintaining a counseling relationship with clients.

Ronnestad and Skovholt (1993) note that beginning students value a supervisor who teaches and provides structure and direction to them. Several researchers, however, emphasize the importance of both the managerial and clinical roles of supervisors. These dual roles serve to support interns in areas such as client selection and provision of relevant agency and client information (Holloway, 1995); monitoring the quality of services provided to clients (Bernard & Goodyear, 1998); providing feedback about performance (Watkins, 1997); and overseeing the professional development of the intern throughout the training experience (Bernard & Goodyear, 1998).

In many ways what follows is our conception of the ideal supervisor, the super supervisor. We like to think of this person as the SUPERvisor. This is certainly unfair to supervisors in that it is not always possible to demonstrate perfection in deed or in person. Nonetheless, for our purposes, we shall demand perfection!

You must feel professionally and personally comfortable with your supervisor and believe that this person will be a good role model and advocate for you. The time you spend together should provide ample opportunities for you to get to know each other as well as to assess your ability to work together. Even though the supervisor maintains major responsibility for the integrity of the internship and for encouraging you as you pursue a career in counseling, your belief that he or she can work with you is vital. Ideally, the supervisor and you share the hope that the internship will be a worthwhile and enjoyable experience for both of you. Any hesitancy you have should be discussed candidly with your site supervisor or university supervisor so any necessary accommodations can be made.

The internship experience is in many ways similar to the counseling relationship; therefore, you may want to work with a supervisor whose therapeutic orientation, ideology, and style are similar to yours. Conversely, however, some interns may benefit from supervisors who differ from them in these areas, since they can be challenged to grow professionally.

One of the key functions of your supervisor is to provide feedback on your performance and progress. This critical feedback process will help you stay focused on both the quality and the quantity of your services. As part of the ACA *Code of Ethics* (Section F), your supervisor accepts the responsibility to help ensure that you develop into a competent counselor. By providing feedback and direction he or she provides a great preparatory service in your professional development.

A supervisor is an educator, and as such he or she enhances your formal academic training. The site supervisor's role modeling may uniquely influence your approach to the profession. Our ideal supervisor assists you in feeling comfortable in awkward or new situations. Moreover, we hope that your supervisor reveals humanity as well as his or her professional persona, and feels comfortable enough to let you know the intricacies of the profession by showing you his or her honest attitudes, emotions, and behavior. You will find his or her support and direction key factors in your continuing journey into the counseling profession. The site supervisor's task is serious; he or she has made a commitment to counseling itself by agreeing to supervise you.

Our ideal supervisor uses such methods of supervision as performing co-therapy with you as intern, observing and critiquing your "live" therapy sessions, and reviewing tapes of sessions in order to provide formal feedback. Supervision is critical to your learning, and good supervision goes beyond signing off on paperwork and talking about courses.

At times you may find that the site supervisor will share facets of the profession that you have not learned about in textbooks. We anticipate that his or her experience, training, and skills will afford you the opportunity to learn much about the field. In essence, he or she might share "the good, the bad, and the ugly" of the counseling field, all of which can be valuable as you continue your professional development. Take all experiences as learning experiences; even the bad can be looked on (or, as some say, reframed) as a challenge and an opportunity.

One issue students often address is how to interact with other mental health professionals at the site. Most likely, your supervisor will ask his or her colleagues to expose you to their professional styles, approaches, and theories. Through such exposure, you may find that their vantage point sometimes differs from your supervisor's. This can pose a dilemma and perhaps be confusing to you. Our ideal supervisor is acutely aware that his or her orientation is not absolute. However, he or she will likely desire that you adhere to his or her directives because ultimate responsibility rests with the supervisor. He or she will be willing to discuss with you any differences you have observed and, generally, will use such experiences as teaching tools. Ultimately, by internal (sometimes unconscious) reflection you will decide how all these bits of information, observations, and various approaches fit you as a therapist. In the interim, you will save yourself much frustration and cognitive dissonance by relying on your supervisor to achieve your educational goals and objectives.

Work closely with your supervisor to make the internship experience an enjoyable, educational, and personally rewarding opportunity. Do not hesitate to share with him or her where your interests lie and what you like and dislike about the total experience, including agency ambience, clients, and staff. Being candid with your supervisor can optimize this experience, enhancing both your learning and your relationship. At the same time, expect candor from your supervisor. Your placement is your testing ground for academic knowledge and should be as pleasant and beneficial as possible for the best learning results. No doubt your supervisor will share this goal with you.

As noted previously, this is probably the final stage of your academic endeavor and, as such, is the pinnacle of a course of study. Among the many thoughts that might occur to you at this point is, "Do I really want to be a counselor?" Don't be overly concerned about such thoughts, but be willing to explore them with your supervisor. Also, integrating your formal education with practice may cause you to realize that the counseling field is not exactly what you thought, were told, or read it would be. Believing something about the profession without validating it does a great disservice to yourself, your colleagues, and, above all, your future clients. Use the placement to clarify your beliefs and attitudes about the profession. Because your site supervisor and university academic adviser have worked with you up to this point to establish your professional goals and objectives, it is most appropriate to discuss this key factor as well.

Often, separating your professional and personal involvement with your supervisor is difficult because the nature of the placement can cause blurred role boundaries. The ideal supervisor adheres closely to the ACA *Code of Ethics and Standards of Practice* (1995) and the Association for Counselor Education and Supervision (ACES) *Ethical Guidelines for Counseling Supervisors* (1993), so his or her personal side can influence the tone, atmosphere, and ethos of the placement. Exposing his or her shortcomings as well as strengths reinforces the

uniqueness and value of your relationship. Further, being a person first enhances the value of the quality of interaction. Of course, supervision is not intended to supplement therapy for the intern. Supervision has as its focus the
continued well-being of the client and the professional growth and development of you as an intern.

It is a rare supervisor who does not learn from his or her students. For example, your current involvement in the academic arena may augment his or
her knowledge base. Your supervisor may take pride in helping to shape and
mold your career—certainly a powerful personal learning experience. The relationship you share can leave a positive and lasting impression on you both.

BEFORE INTERNSHIP

Now let's look at this special relationship as it develops from beginning to end.
A number of tasks need to be achieved when you first meet your supervisor.
He or she will need to know your university and state licensing or certification
requirements as well as your personal goals. The placement can influence your
belief about the counseling profession itself, so it may be helpful for the supervisor to require definitive, objective, and measurable goals. Additionally, he or
she has the responsibility to understand who you are, especially as an aspiring
counseling professional. Because the supervisor is aware of the placement site's
policies and procedures, he or she will gauge the appropriateness of a match
and the potential for successful completion. It is not unusual for the supervisor
to conduct a somewhat formal interview with you to learn why you chose this
site, what your career objectives are, and what you would like to achieve in
your professional relationship with him or her. Your level of understanding regarding theories of counseling, techniques, and the counseling field in general
will also be important. As a professional "gate keeper," the supervisor will assess
whether you will be able to integrate your book knowledge from your academic training at this specific site. In addition to examining your academic
foundation, your supervisor will also consider your personal attributes.

No two training sites have the same mix of personnel, client population,
and agency standards and ethos. For these reasons, the supervisor, as presumed
expert of his or her professional domain, must make the final decision regarding your acceptance. Once you are accepted, however, our ideal supervisor will
outline for you any forms you need to complete before starting the placement.
You may be fingerprinted if working with children. You may also be expected
to provide evidence of liability insurance coverage. The supervisor may also
suggest specific readings, including pertinent agency orientation materials such
as policies and procedures. Perhaps you will be asked to review your theories
and techniques as well as to start solidifying a therapeutic orientation based on
your knowledge of the clinical population. You may also be asked to prepare
audiotape or videotape role-play demonstrations so the supervisor can establish
a baseline of your knowledge and skill level and which could be reviewed in

group supervision settings. The supervisor may discuss with you how you feel personally about working with certain clinical populations to see if any personal issues may impede your ability to be effective and efficient with clients. Professional involvement will not be devoid of the personal feelings, attitudes, and beliefs that are a part of who you are.

As the final preparations are made for your placement, the supervisor will explain in depth the agency's policies, procedures, mission statements, and other information relevant to your placement. Specific attention may be given to past involvement of students at this site and how they fit into the overall functioning of the organization. The supervisor will help you to understand how you will be perceived during your placement, what specific parameters will be in place, and how you will fit into the organization. Ideally, your supervisor will be an advocate for you and has agreed to shoulder the responsibility to have you integrate successfully into the treatment milieu and the organizational structure.

The routine and seemingly mundane tasks of completing the necessary academic course forms (including registration) and student/supervisor agreements should be done at your earliest convenience. Watch for university and agency deadlines! If you are governed by specific licensure and/or certification requirements, understanding these standards and promptly filling out any necessary forms are vital to receiving proper experiential credit. Failure to do so could result in a delay in your ability to see clients.

DURING INTERNSHIP

Internship is a great opportunity to put your skills and knowledge to use, and our exemplary supervisor will provide you with the chance to use them to the fullest. The ideal is to have you experience all the clinical aspects possible within the organization. What better way to delve into the "real world" of counseling than to practice one's skills in all areas of clinical services?

Barring any sensitive issues for which clients may prefer not to have you present, it is appropriate and a valuable learning experience to "shadow" the supervisor (or other clinical staff) in all his or her clinical activities. By observing the supervisor and eventually participating in services, you will gradually become acclimated to the setting and begin integrating your book knowledge with clinical practice. Shadowing lets both of you assess the scope of your knowledge and level of comfort as you aspire to independent work with clients. The decision to work independently with clients (still under supervision) will be determined by you and your supervisor based on your "readiness" to accept this responsibility. Further, our clients unfortunately do not always fit into our plans as to when and what types of techniques or approaches we would like to practice with them. You should feel comfortable discussing with your supervisor your perceived readiness to accept more responsibility and to use your knowledge and skills. Take note of how it feels to use your knowledge.

One critical issue to address throughout the internship is what treatment approach to use with your clients. Before you solidify your approach, review with your supervisor your philosophy of humankind (see Chapter 2 on self-assessment). Perhaps you have previously dealt with this personal philosophical stance in your theoretical studies, but now you will link those theories with your belief system in a practical setting with contact with fellow human beings, who are, for the most part, suffering. We believe that your philosophy of humankind affects the way you approach and treat your clients, and, for that matter, people in general. For many students this will be the first time they closely examine how their beliefs can influence their dealings with clients in vivo. Once you begin to formulate, review, and refine your philosophy of humankind, our ideal supervisor can guide you in selecting a philosophically and theoretically congruent treatment approach. For example, existential approaches lend themselves more to freedom of choice, whereas analytically and behaviorally oriented approaches appear to be deterministic.

Throughout the internship your supervisor will ideally schedule both formal and informal supervision conferences with you. During these meetings you should review your internship goals and objectives to assess adherence, progress, and any necessary modifications to them. Your supervisor will give you constructive feedback as to the quality of your performance, including recommendations for improvement. You will generally be formally evaluated midway through the placement and again at the end; your academic adviser may be part of this process as well. Particular attention may be given to how well you feel you have become part of the agency's treatment structure and the overall treatment team; your supervisor will validate your perceptions of these two areas. In addition, your supervisor may help you consider whether a specific clinical population or treatment specialty may be indicated and elaborate on any areas in which you are demonstrating competency (see Chapter 4, which discusses competencies you should develop). Remember, too, that you also must take responsibility for making your experience a good one. Once you appreciate the agency milieu and get a sense of agency operations, we suggest that—when you have a particular need—you ask, and not demand, that the need be met. Some interns make efforts to do informational interviews with staff, tag along on professional lunches, or read material related to specific clients served, for example.

AT THE TERMINATION OF INTERNSHIP

Your site supervisor has served in a highly significant role and has maintained a considerable amount of responsibility throughout your internship. One of his or her most enjoyable tasks is to help you bring closure to your placement experience. Your supervisor has worked closely with you by investing time, knowledge, and skill to help you become a competent counselor-in-training. He or she will be eager to share impressions and recommendations with you as to your clinical competencies, skills, and knowledge base. The supervisor will provide feedback on how well you were able to translate your academic foundation to the practical work situation. His or her input, guidance, and direction can influence your career focus. This is a responsibility in which the supervisor takes pride. The completion of your final evaluation, often in concert with your academic adviser, will formally appraise your efforts. This formal document presents measurable, objective, and observable data about your clinical strengths and weaknesses, general impressions, and overall recommendations.

You should have ample opportunity to respond to the evaluation and, if necessary, get clarification on its content. Further, you may be invited to evaluate both your site and supervisor in what we view as a reciprocal process. (Appendices M and N present sample evaluation forms of the intern and the site and supervisor, respectively.) Who knows? If all goes well, you may be invited to join the clinical staff!

A WORD ABOUT YOUR
UNIVERSITY SUPERVISOR

We assume that by the time you get to internship placement you have formed a relationship with your university supervisor. Perhaps you have had him or her for classes previously; perhaps the university supervisor serves as your program adviser. In any case, under most circumstances this person serves as the university liaison between your agency and the professor of the integrating internship course on campus. We strongly suggest that you take any and all placement problems to your university supervisor early on to avoid frustration and consternation.

Most programs offer a weekly or biweekly seminar on campus, conducted by the university supervisor, during which interns process their experiences and present and critique cases (masking any identifying client data, of course). You are likely to be formally evaluated by the university supervisor, too. The university supervisor usually assigns the course grade, generally in consultation with the site supervisor.

Finally, whenever you meet with your university supervisor—and with your site supervisor as well—always be on time and be prepared!

REFERENCES

ASSOCIATION FOR COUNSELOR EDUCATION AND SUPERVISION. (1993). *Ethical guidelines for counseling supervisors.* Available at: www.siu.edu/~epse1/aces/documents/ethicsnoframe.htm

BERNARD, J. M., & GOODYEAR, R. K. (1998). *Fundamentals of clinical supervision* (2nd ed.). Boston: Allyn & Bacon.

HOLLOWAY, E. (1995). *Clinical supervision: A systems approach.* Thousand Oaks, CA: Sage.

RONNESTAD, M., & SKOVHOLT, T. (1993). Supervision of beginning and advanced graduate students of counseling and psychotherapy. *Journal of Counseling and Development, 71*(4), 396–405.

WATKINS, C. E., Jr. (1997). Defining psychotherapy supervision and under-

standing supervisor functioning. In C. E. Watkins, Jr. (Ed.), *Handbook of psychotherapy supervision* (pp. 3–10). New York: John Wiley & Sons.

BIBLIOGRAPHY

ANCIS, J. R., & LADANY, N. (2001). A multicultural framework for counselor supervision. In L. J. Bradley & N. Ladany (Eds.), *Counselor supervision: Principles, process, and practice* (3rd ed., pp. 63–90). Philadelphia: Brunner-Routledge.

BERNARD, J. M., & GOODYEAR, R. K. (1998). *Fundamentals of clinical supervision* (2nd ed.). Boston: Allyn & Bacon.

BRADLEY, L. J., & GOULD, L. J. (2001). Psychotherapy-based models of counselor supervision. In L. J. Bradley & N. Ladany (Eds.), *Counselor supervision: Principles, process, and practice* (3rd ed., pp. 147–180). Philadelphia: Brunner-Routledge.

BRADLEY, L. J., & LADANY, N. (Eds.). (2001). *Counselor supervision: Principles, process, and practice* (3rd ed.). Philadelphia: Brunner-Routledge.

BRADLEY, L. J., & PLANNY, K. J. (2001). Supervision-based developmental models of counselor supervision. In L. J. Bradley & N. Ladany (Eds.), *Counselor supervision: Principles, process, and practice* (3rd ed., pp. 125–146). Philadelphia: Brunner-Routledge.

BRADLEY, L. J., & WHITING, P. P. (2001). Supervision training: A model. In L. J. Bradley & N. Ladany (Eds.), *Counselor supervision: Principles, process, and practice* (3rd ed., pp. 361–387). Philadelphia: Brunner-Routledge.

CHIAFERI, R., & GRIFFIN, M. (1997). *Developing fieldwork skills: A guide for human services, counseling, and social work students.* Pacific Grove, CA: Brooks/Cole.

COOK, D. A., & HELMS, J. E. (1998). Visible racial/ethnic group super-

visees' satisfaction with cross-cultural supervision as predicted by relationship characteristics. *Journal of Counseling Psychology, 35,* 268–274.

EDWARDS, J. K., & CHEN, M. W. (1999). Strength-based supervision: Frameworks, current practice, and future directions—a Wu-wei method. *Family Journal: Counseling and Therapy for Couples and Families, 7,* 349–357.

GARRETT, M. T., BORDERS, L. D., CRUTCHFIELD, L. B., TORRES-RIVERA, E., BROTHERTON, D., & CURTIS, R. (2001). Multicultural superVISION: A paradigm of cultural responsiveness for supervisors. *Journal of Multicultural Counseling and Development, 29,* 147–158.

GUEST, C. L., Jr., & DOOLEY, K. (1999). Supervisor malpractice: Liability to the supervisee in clinical supervision. *Counselor Education and Supervision, 38,* 269–279.

HELMS, J. E., & COOK, D. A. (1999). *Using race and culture in counseling and psychotherapy: Theory and process.* Boston: Allyn & Bacon.

HESS, A. (1980). *Psychotherapy supervision: Theory, research and practice.* New York: Wiley.

HOLLOWAY, E. (1995). *Clinical supervision: A systems approach.* Thousand Oaks, CA: Sage.

JORDAN, K. (1998). The cultural experiences and identified needs of the ethnic minority supervisee in the context of Caucasian supervision. *Family Therapy, 25,* 181–187.

KAISER, T. (1997). *Supervisory relationships: Exploring the human element.* Pacific Grove, CA: Brooks/Cole.

LADANY, N., ELLIS, M. V., & FRIEDLANDER, M. L. (1999). The supervisory working alliance, trainee self-efficacy, and satisfaction. *Journal of Counseling and Development, 77,* 447–455.

LEE, R. W., & GILLAM, S. L. (2000). Legal and ethical issues involving the duty to warn: Implications for supervisors. *Clinical Supervisor, 19,* 123–136.

MURATORI, M. C. (2001). Examining supervisor impairment from the counselor trainee's perspective. *Counselor Education and Supervision, 41,* 41–56.

NEUFELDT, S. (1999). *Supervision strategies for the first practicum* (2nd ed.). Alexandria, VA: American Counseling Association.

NEUKRUG, E. S. (1998). *The world of the counselor: An introduction to the counseling profession.* Pacific Grove, CA: Wadsworth.

POPE-DAVIS, D., & COLEMAN, H. L. K. (Eds.). (1997). *Multicultural counseling competencies: Assessment, education, training and supervision.* Thousand Oaks, CA: Sage.

STOLTENBERG, C., MCNEILL, B., & DELWORTH, U. (1998). *IDM supervision: An integrated developmental model for supervision counselors and therapists.* San Francisco: Jossey-Bass.

SUE, D. W., & SUE, D. (1999). *Counseling the culturally different: Theory and practice.* New York: John Wiley & Sons.

WONG, P. T. P., & WONG, C. J. (1999). Assessing multicultural supervision competencies. In W. J. Lonner & D. L. Dinnel (Eds.), *Merging past, present and future in cross-cultural psychology: Selected papers from the 14th interactive Congress of the International Association of Cross-cultural Psychology* (pp. 510–519). Lisse, Netherlands: Swetszeitlinger.

4

Developing Competencies and Demonstrating Skills

Your internship provides an arena for you to try your wings as a helping professional with guidance and support close at hand. Many students feel a bit overwhelmed as they begin to interact with clients and staff members. You may have difficulty at first as you try to remember counseling theory and techniques, to recall academic coursework concerning such areas as human development or multicultural issues, to keep in mind ethical guidelines, and to think about agency procedures, regulations, and policies—all while trying to attend to your first few clients! More than one intern has felt discouraged after the first week of trying to juggle all the responsibilities of the new role.

Your internship can be viewed as a time to build a framework of new professional relational skills on a foundation of the material you have learned in your counseling program courses, your own life experiences, and your personal values and philosophies. This framework is composed of new perspectives, understandings, abilities, and skills added gradually and with care. Your goal is to construct a strong framework over a solid foundation, working diligently but patiently, and often standing back to look at the work you have accomplished so far.

During your internship you will develop some of the specific personal attributes and competencies that you will use during your professional counseling career. To help you delineate your goals, we have compiled the following list of skills for graduate-level internship students to work toward building. In reality, not all placement sites afford the opportunity to develop abilities in every area we have indicated. In addition, the quantity and scope of the competencies listed here reflect our belief that becoming a professional counselor is an ongoing process. Stoltenberg, McNeill, and Delworth (1998) write that supervision is an interactive process that does not necessarily follow a continuous, teleological (step-by-step) course, but varies as new tasks, settings, or other challenges emerge over the course of time. Your internship is just the beginning of your professional development; you will continue to add competencies throughout your career.

SUGGESTED COMPETENCIES FOR INTERNS

I. Communication Skills
   A. Verbal skills
      1. Students will be able to express themselves clearly and concisely in daily interactions with agency staff members and other professionals.
      2. Students will be able to communicate pertinent information about clients and to participate effectively in interdisciplinary treatment team meetings and case conferences (including case presentations, which may involve videotaping and/or audiotaping), while maintaining their identities as counselors within a multidisciplinary group.
      3. Students will be able to educate clients and to provide appropriate information on a variety of issues (such as parenting, after-care and other support services, psychotropic medications, stress management, sexuality, or psychiatric disorders) in an easily understandable manner.
      4. Students will be able to communicate with clients' families, significant others, and designated friends in a helpful fashion. They will be able to provide, as well as to obtain, information concerning the client, while respecting the client's rights concerning privacy, confidentiality, and informed consent.
      5. Students will be able to communicate effectively with referral sources, both inside and outside the agency, concerning all aspects of client needs and well-being (for example, housing, legal issues, healthcare, Twelve Step programs, and psychiatric concerns).
   B. Writing skills
      1. Students will be able to prepare a complete, written initial intake assessment, including a mental status evaluation, a psychosocial history, a diagnostic impression, and recommended treatment modalities.
      2. Students will be able to write progress notes, to chart, and to maintain client records according to agency standards and regulations.
      3. Students will be able to prepare a written treatment plan, including client problems, therapeutic goals, and specific interventions to

be utilized. This plan is concrete, behaviorally specific, and individualized to the client.

4. Students will be able to prepare and present a formal, written case study.
5. Students will be able to use computer skills to work with word processing programs and to maintain and search databases.

C. Knowledge of nomenclature
1. Students will thoroughly know professional terminology pertaining to counseling, psychopathology, treatment modalities, and psychotropic medications.
2. Students will be able to understand professional counseling jargon and will be able to participate in professional dialogue.

II. Interviewing
A. Students will structure the interview according to a specific theoretical perspective (for example, psychodynamic or behavioral theory) because a theory base provides the framework and rationale for all therapeutic strategies, techniques, and interventions.
B. Students will be able to use appropriate counseling techniques to engage the client in the interviewing process, to build and maintain rapport, and to begin to establish a therapeutic alliance. This may include using attending behaviors, active listening skills, and a knowledgeable and professional attitude to convey empathy, genuineness, respect, and caring, and to be perceived as trustworthy, competent, helpful, and expert (Cormier & Cormier, 1998).
C. Students will be able to use appropriate counseling techniques to increase client comfort and to facilitate collection of data necessary for clinical assessment, such as evaluating mental status, taking a thorough psychosocial history, and eliciting relevant, valid information concerning the presenting problem, in order to formulate a diagnostic impression. Specific interviewing competencies may include observation, use of open-ended and closed-ended questions, the ability to help the client stay focused, reflection of content and feeling, reassuring and supportive interventions, and the ability to convey an accepting and nonjudgmental attitude.
D. Students will develop a holistic approach toward interviewing by assessing psychological, biological, environmental, and interpersonal factors that may have contributed to the client's developmental history and presenting problems.
E. Students will strive to see things from the client's frame of reference and to develop a growing understanding of the client's phenomenological perspective.
F. Students will be aware at all times of the crucial importance of understanding the client from a multicultural perspective and will be aware that sociocultural heritage is a key factor in determining the client's unique sense of self, worldview, values, ideals, patterns of interpersonal communication, spiritual/religious views, family structure, behavioral norms, and concepts of wellness as well as of pathology.

III. Diagnosis

  A. Students will understand the most commonly used assessment instruments, such as personality and intelligence tests, anxiety and depression scales, and interest inventories.

    1. Students will become familiar with the validity and reliability of these instruments.

    2. Students will be able to interpret data generated by these instruments and understand the significance of these data in relation to diagnosis and treatment.

    3. Students will be able to determine which assessment instruments would be most helpful in evaluating specific client problems or concerns.

    4. Students will be aware of the limitations of assessment instruments when used with ethnic minority populations.

  B. Students will develop a working knowledge of the *Diagnostic and Statistical Manual of Mental Disorders* (DSM-IV-TR).

    1. Students will be familiar with the organization of the DSM-IV-TR and will be able to use this nosology effectively (for example, to find diagnostic codes or to trace clients' behaviors, affects, or cognitions along the decision trees to ascertain potential diagnoses).

    2. Students will be able to understand the DSM-IV-TR classification of disorders and will be able to identify particular constellations of client problems as specific DSM-IV-TR diagnostic categories.

  C. Students will be able to review and consider all pertinent data, including interviews, medical records, previous psychiatric records, test results, psychosocial history, consultations, and DSM-IV-TR classifications, in formulating a diagnostic impression or preliminary diagnosis.

IV. Treatment

  A. Students will be able to conduct therapy using accepted and appropriate treatment modalities and counseling techniques based on recognized theoretical orientations and outcome research.

    1. Students will work toward identifying their own theoretical frameworks based on their own philosophy of humankind.

    2. Students will know how to make treatment recommendations, formulate a treatment plan, establish a treatment contract, implement therapy, and terminate the therapeutic relationship at an appropriate time. (Refer to the Sample Treatment Plan in Appendix K.)

    3. Students will be able to conduct the following types of therapy and will understand the underlying principles, issues, dynamics, and role of the counselor associated with each type of treatment or treatment-related activity:

      a. Conjoint therapy

      b. Crisis intervention

  c. Family therapy
  d. Group therapy
  e. Individual therapy
  f. Marital/couples therapy
  g. Brief models of therapy
  h. Play therapy
  i. Mental health consultation
B. Students will understand that different client populations and different types of problems may respond best to varying therapeutic approaches and techniques.
  1. Students will be knowledgeable about various types of client populations and their particular problems and concerns, including but not limited to the following:
    a. Adult children of alcoholics
    b. Adults
    c. Chemically dependent individuals
    d. Children and adolescents
    e. Clients of varied ethnic, cultural, and religious backgrounds
    f. Dual-diagnosed clients (for example, chemically dependent with a psychiatric disorder)
    g. Individuals with eating disorders
    h. Elders
    i. Gay, lesbian, bisexual, and transgender clients
    j. Physically or cognitively impaired clients
    k. Survivors of trauma
    l. Any other individual who may be included under the Americans with Disabilities Act
  2. Students will be flexible and knowledgeable in determining population-appropriate counseling techniques and therapeutic interventions. Students will have as many therapeutic tools available for use as possible (for example, play therapy, art therapy, behavioral techniques, role-playing, Gestalt techniques, directive versus nondirective techniques, stress management techniques, experiential therapy, and hypnosis).
C. Students will be sensitive to the impact of multicultural issues and diversity on the counseling relationship and on treatment, and will modify therapeutic approaches and techniques to respect multicultural differences and to meet multicultural needs.
D. Students will be able to direct clients to appropriate sources of information, such as books, Web sites, and so forth.

V. Case Management
A. Students will understand the functions and goals of all departments, programs, and services within the agency and will be able to network with appropriate personnel throughout the social service system.
B. Students will understand the roles, responsibilities, and contributions to client care of members in each department or program within the

agency. The student will know which individual(s) to contact to help resolve various client problems.

    C. Students will acquire a thorough knowledge of community resources and will understand the agency procedures for referring clients to outside sources for help.

    D. Students will consider continuity of care to be a most important goal, beginning with the initial client contact.

        1. Students will act as an advocate for the client in ensuring continued quality of care and access to social services. Advocacy will include, but not be limited to, exploring possible funding sources for care, such as mental health coverage on insurance policies, Medicaid, or Medicare.

        2. Students will collaborate with other agencies or institutions, which also serve the client.

        3. Students will be able to participate in all areas of discharge planning, including arranging follow-up visits with a mental health professional, communicating with insurance companies, and providing help with housing, transportation, vocational guidance, legal assistance, support groups, medical care, and referral to other services or agencies.

VI. Agency Operations and Administration

    A. Students will be familiar with the organizational structure, including the table of organization of the agency, and will understand the responsibilities and functions of administrative staff.

    B. Students will understand the philosophy, mission, and goals of the agency and will thoroughly know the agency's policies and procedures, which are usually delineated in a comprehensive manual.

    C. Students should be aware of immediate and long-range strategic plans for the agency (for example, to hire an art therapist, to develop a chemical abuse program, or to add an additional building, as well as to evaluate and eliminate ineffective programs).

    D. Students will understand the business aspects of the agency (for example, funding sources and managed care budget allowances). Productivity is the catch word!

    E. Students will be aware of legal issues concerning agency functions, such as state or national licensure/certification requirements or safety regulations.

    F. Students will understand agency standards that ensure continued quality of care, including quality assurance and peer review processes.

    G. Students will avail themselves of the latest technology in order to better assist clients.

VII. Professional Orientation

    A. Students will know all ethical and legal codes for counselors, provided by professional counseling associations as well as by state law, and will adhere to these standards at all times.

B. Students will be familiar with agency regulations and policies regarding ethical and legal issues and will adhere to these standards at the placement site.
C. Students will be knowledgeable concerning legislation protecting human rights.
D. Students will seek guidance from the on-site supervisor and the academic program supervisor with any questions concerning ethical or legal issues or professional behavior.
E. Students will consider the four basic R's for counselors (Carkhuff, 1993) whenever acting in a professional helping capacity: the right of the counselor to intervene in the client's life, the responsibility the counselor assumes when intervening, the special role the counselor plays in the helping process, and the realization of the counselor's own resources in being helpful to the client.

You may wish to look at the Sample Evaluation of Intern Form in Appendix M as well as Appendix N's Sample Intern Evaluation of Site/Supervisor Form. These forms list the competencies addressed in this chapter for rating purposes.

REFERENCES

CARKHUFF, R. R. (1993). *The art of helping* (7th ed.). Amherst, MA: Human Resource Development Press.

CORMIER, S., & CORMIER, B. (1998). *Interviewing strategies for helpers* (4th ed.). Pacific Grove, CA: Brooks/Cole.

STOLTENBERG, C. D., MCNEILL, B., & DELWORTH, U. (1998). *IDM supervision: An integrated developmental model for supervising counselors and therapists.* San Francisco: Jossey-Bass.

BIBLIOGRAPHY

AMERICAN PSYCHIATRIC ASSOCIATION. (2000). *Diagnostic and statistical manual of mental disorders* (4th ed.) (Text revision). Washington, DC: Author.

AXLINE, V. M. (1993). *Play therapy* (Rev. ed.). New York: Ballantine.

BALSAM, R. M., & BALSAM, A. (1984). *Becoming a psychotherapist: A clinical primer* (2nd ed.). Chicago: University of Chicago Press.

BALTIMORE, M. L. (2000). Ethical considerations in the use of technology for marriage and family counselors. *The Family Journal: Counseling and Therapy for Couples and Families, 8,* 390–393.

BENJAMIN, A. (1987). *The helping interview with case illustrations.* Boston: Houghton Mifflin.

CASEY, J. A., BLOOM, J. W., & MOAN, E. R. (1994). Use of technology in counselor supervision. *ERIC Clearinghouse on Counseling and Student Services.* ED372357.

CHRISTIE, B. S. *Counseling supervisees of distance clinical supervision.* Available at: cybercounsel.uncg.edu/book/manuscripts/tenets.htm

COURSOL, D. H., & LEWIS, J. *Cyber-supervision: Close encounters in the new millennium.* Available at cybercounsel.uncg.edu/book/manuscripts/tenets.htm

DAVIS, D. C., & HUMPHREY, K. M. (Eds.). (2000). *College counseling: Issues and strategies for a new millennium.* Alexandria, VA: American Counseling Association.

FAIVER, C., INGERSOLL, E., O'BRIEN, E., & McNALLY, C. 2001). *Explorations in counseling and spirituality: Philosophical, practical, and personal reflections.* Pacific Grove, CA: Brooks/Cole.

GETZ, H. G., & SCHNUMAN-CROOK, A. (2001). Utilization of online training for on-site clinical supervision. *Journal of Technology in Counseling, 2.1.* Available at: jtc.colstate.edu/Vol2_1/Supervisors.htm

GROHOL, J. M. (1997). *Starting a new online support group.* Available at: www.grohol.com

HART, J. L. *Mentoring without walls: Using cyberspace to enhance student-faculty guidance.* Available at: cybercounsel. uncg.edu/book/manuscripts/tenets.htm

HINTERKOPF, E. (1998). *Integrating spirituality into counseling: A manual for using the experiential focusing method.* Alexandria, VA: American Counseling Association.

HOHENSHIL, T. H. (2000). High tech counseling. *Journal of Counseling and Development, 78,* 365–368.

HOOD, A. B., & JOHNSON, R. W. (1997). *Assessment in counseling: A guide to the use of psychological assessment procedures* (2nd ed.). Alexandria, VA: American Counseling Association.

JANOFF, D. S., & SCHOENOLTZ-READ, J. (1999). Group supervision meets technology: A model for computer-mediated group training at a distance. *International Journal of Group Psychotherapy, 49,* 255–271.

KOTTMAN, T. (2001). *Play therapy: Basics and beyond.* Alexandria, VA: American Counseling Association.

LEE, C. C. (2002). *Multicultural issues in counseling: New approaches to diversity* (3rd ed.). Alexandria, VA: American Counseling Association.

LEWIS, J., COURSEL, D., KHAN, L., & WILSON, A. *Life in a dot.com world: Preparing counselors to work with technology.* Available at: cybercounsel.uncg.edu/book/manuscripts/tenets.htm

MCBRIDE, J. L. (1998). *Spiritual crisis: Surviving trauma to the soul.* New York: Haworth Pastoral Press.

MILLER, W. R. (Ed.). (1999). *Integrating spirituality into treatment: Resources for practitioners.* Washington, DC: American Psychological Association.

MITCHELL, R. (2001). *Documentation in counseling records* (2nd ed.). Alexandria, VA: American Counseling Association.

MYERS, J. E., & GIBSON, D. M. (1999). Technology competence of counselor educators. *ERIC Clearinghouse on Counseling and Student Services.* ED00036.

NATHAN, P. E., & GORMAN, J. M. (Eds.). (1998). *A guide to treatments that work.* New York: Oxford University Press.

ORAVEC, J. A. (2000). Online counseling and the Internet: Perspectives for mental health care supervision and education. *Journal of Mental Health, 9,* 121–135.

PEDERSON, P. (2000). *A handbook for developing multicultural awareness* (3rd ed.). Alexandria, VA: American Counseling Association.

REID, W. H., & WISE, M. G. (1995). *DSM-IV training guide.* New York: Brunner/Mazel.

RICHARDS, P. S., & BERGIN, A. E. (1997). *A spiritual strategy: For counseling and psychotherapy.* Washington, DC: American Psychological Association.

RICHARDS, P. S., & BERGIN, A. E. (Eds.). (2000). *Handbook of psychotherapy and religious diversity.* Washington, DC: American Psychological Association.

SCHERL, C. R., & HALEY, J. (2000). Computer monitor supervision: A clinical note. *The American Journal of Family Therapy, 28,* 275–282.

SEXTON, T. L., WHISTON, S. C., BLEUER, J. C., & WALZ, G. R. (1997). *Integrating outcome research into counseling practice and training.* Alexandria, VA: American Counseling Association.

SHEA, S. (1998). *Psychiatric interviewing: The art of understanding* (2nd ed.). Philadelphia: Saunders.

SPITZER, R. L., GIBBON, M., SKODOL, A., WILLIAMS, J., & FIRST, M. (Eds.). (1997). *DSM-IV casebook.* Washington, DC: American Psychiatric Press.

WALLACE, S., & LEWIS, M. (1998). *Becoming a professional counselor: Preparing for certification and comprehensive exams* (2nd ed.). London: Sage.

WELFEL, E. R., & HEINLEN, K. T. (2001). The responsible use of technology in mental health practice. In E. R. Welfel & R. E. Ingersoll (Eds.), *The mental health desk reference* (pp. 484–490). New York: John Wiley & Sons.

YALOM, I. (1995). *Theory and practice of group psychotherapy* (4th ed.). New York: Basic Books.

5

The Clinical Interview

Many interns begin their experience by performing a clinical interview as part of the intake process. Note our intentional use of the word *process*. We view the intake as an important component of the therapy itself, rather than as a distinct piece of "busy work" apart from the dynamics of counseling. In other words, the clinical interview starts establishing the therapeutic relationship between the counselor and client, which conveys a sense of client worth and worthiness for treatment. It has been the authors' experience that in many agencies and institutions, clinicians view the intake as tedious, somewhat routine, and requiring minimal skills. As a result, interns or novice clinicians often perform the intake. Generally, the intake is the first contact many clients have with the mental health system and with therapy and counseling itself; therefore, it sets the tone for treatment, for sharing agency and counselor philosophy and values, and for treatment expectations.

The many types and methods of intake depend mostly on agency policy and style. Some are rather informal, calling for minimal information. Others are lengthy, formalized procedures of data gathering involving the completion of several intricate forms, some by the client, others by the counselor. Still others make use of testing and inventories (Dougherty & Chamblin, 1999; Fisher, Beutler, & Williams, 1999; Maruish, 1999), while some offer interviews specific to dysfunctions and diagnosis (Degood, Crawford, & Jongsma, 1999; Frances & Ross, 1996; Jensen, 1999; Kadden & Skerker, 1999; King & Scahill, 1999; Klassen et al., 1999; Othmer & Othmer, 1994; Paleg & Jongsma, 2000;

Seligman, 1990; and Spitzer, Gibbon, Skodol, & First, 1994). Finally, some agencies employ workers whose job is solely that of performing intake assessments. All share in the goal of helping clients successfully empower themselves to address presenting problems. Refer to Appendices J and K for sample intake and treatment plan forms, respectively.

We view the intake as a systematic process of information gathering and sharing that usually includes the following areas: (1) client description, (2) problem description, (3) psychosocial history, (4) mental status examination, (5) diagnostic impression, and (6) treatment recommendations and initial plan (Faiver, 2001). Further, as Benjamin (1987) notes, the counselor-in-training should view the counseling relationship as special and unique, thereby ensuring full concentration of energy on the client and his or her situation and unique personhood.

The client should be encouraged to ask questions and share expectations regarding treatment. Also, we suggest that you consider discussing what counseling and therapy are and are not. For example, counseling and therapy are a mutual process of personal exploration, growth, hard work, and anticipated problem resolution, not exercises in magical advice-giving and unilateral direction by the counselor. Further, counselors should allow sufficient time to conduct the interview based upon the presenting problem, client's history, level of present functioning, and the client's willingness to engage in this process. Thus, conducting a comprehensive clinical interview as a foundation for successive therapeutic interventions may take more than one session.

At this juncture, let's describe in detail each of the areas of the formal intake process. We suggest asking your supervisor if you can observe him or her (or another clinical staff member who does intakes) one or two times before you attempt the intake yourself.

CLIENT DESCRIPTION

Acquiring basic background information on your client sets the initial phase of the clinical interview. With these data, you establish a general framework by which you initiate a relationship with the client. You will soon realize this knowledge fits into your understanding of key developmental influences of the client's life. We suggest the following information be gathered:

- Name
- Address
- Telephone number
- Emergency contact person with telephone number
- Guardian and/or responsible party involvement for minors
- Birth date

- Gender
- Race
- Nationality
- Marital status/significant other/sexual orientation
- Children
- Religion
- Level of education (including regular or special)
- Financial data (including Medicaid, Medicare, self-pay, or insurance information)
- Socioeconomic information
- Cultural influences
- Present job/position
- Referral source

Once this information is collected, you may already have formulated some beliefs about the individual and the direction to take in counseling.

PROBLEM DESCRIPTION

You must encourage your client to define his or her problem or concern at that particular time and be aware that the explanation of this problem area is from his or her sole vantage point. As the interview progresses, you may discover incongruities or discrepancies that may need sorting out at some point. Delineating the scope of the problem and its effects on the client's and significant others' present levels of functioning is most relevant and a key to the flow of counseling. Inquire about the client's perception of the etiology of the problem, such as when the problem started and any influencing interpersonal and environmental factors. Even though clients present a specific uniqueness to their problems and are unique individuals in and of themselves, you need to convey some assurance that you have dealt with similar concerns, maintaining both your clinical objectivity and respect for the clients' individuality. Clients usually expect that you as "expert" (see Cormier & Cormier, 1998) will be able to state your understanding of the problem area (Othmer & Othmer, 1994). This statement reassures clients and enhances the development of the therapeutic alliance and subsequent therapeutic journey; it also can facilitate the direction of treatment. If you are not sure exactly what the defined problem is, ask and get clarification so as not to impede the direction of the interview. Frequently, the presenting problem may mask a significant underlying issue, which may not surface until trust is established. Further, you can explore the history of similar problems and how they were resolved.

PSYCHOSOCIAL HISTORY

Now that the problem area has been designated, explore the client's background by taking the psychosocial history, which is, in essence, a brief history of important life events of the client rather than an extensive and detailed life history. As such, your psychosocial history should address the developmental milestones to the present.

We follow this format. As with any outline, it should be adjusted to meet your needs.

- Developmental history—emphasizing milestones and anomalies of client development
- Description of client's family—mother, father, brothers, sisters, and any significant others; description of client's relationships with these individuals
- Description of early home life (happy, abusive, or the like)
- Educational history
- Occupational history
- Social activities—dating, sexual history, relationship history, and problems
- Relationship/marital history—including separations, divorces, and significant relationships
- Health history—physical, mental, accidents, illnesses, surgeries, hospitalizations, psychiatric medication (past and present), other medications (past and present), date of last physical examination, and any current medical conditions under treatment
- Substance use, abuse, and dependence
- Legal history—including probation and parole
- Present living arrangements
- Significant life events—including high and low points
- Abuse issues—physical, emotional, and sexual
- Psychiatric history—inpatient and outpatient; medications
- Individuals who have greatly influenced the client's life
- Individuals who influence the client presently
- Past and present support systems (such as parents, friends, or spouse)
- Ethnic/cultural influences
- Religious history and present belief system
- Community involvement, interests, and leisure activities
- Military history—including type of discharge
- Significant environmental influences (such as school, political, and religious)
- Age-specific issues in minors
- Anything else we should know but failed to ask

We suggest that you explore in depth any specific area of history that has the most relevance to the presenting problem and areas of client functioning.

MENTAL STATUS EXAMINATION

The mental status examination is your stethoscope for understanding the client's behavioral, cognitive, and affective domains. In fact, Othmer and Othmer (1994) note that the mental status examination is your reading of the client's "signals," which present ". . . you the cross-sectional view of . . . strengths, weaknesses, and dysfunctions" (p. 100). This evaluation is a key factor in determining a diagnostic impression.

You may want to intersperse the various components of the mental status evaluation throughout your interview. This approach tends to be more comfortable for clients because it avoids the drudgery of a long list of questions.

Faiver (2001) focuses his mental status evaluation on assessing (1) behavior, (2) affect, and (3) cognition. We advise the intern to place equal emphasis on all three areas but to feel comfortable, when necessary, delving into any of them for further clarification. In addition, the order of assessment should be based on personal preference and style. In the next sections, we will focus briefly on each area.

Behavior

Evaluate how the client's behavior affects present functioning. Note overall appearance, including specific physical characteristics, unusual clothing, mannerisms, gestures, tics, motor activities, posture, level of self-care, level of activity, sleeping and eating disturbances, fatigue symptoms, physical or verbal aggression, and the client's reaction to you as therapist. What does the client's body language convey to you? Does his or her posture denote an open or closed stance? Are there any psychosomatic issues? Observe whether the client can attain and maintain eye contact. Note any changes of facial expression or any other noteworthy behaviors. How are the flow and content of speech? Is speech pressured, impoverished, or tangential?

Affect

Next, concern yourself with the client's feelings. Evaluate the client's overall pervasive mood. Is the client's overall feeling level dysphoric (depressed or irritable), expansive, elevated, or euphoric? Note the degree of affect. Is the affect range full and appropriately reactive? Are the client's feeling levels mild, moderate, or severe? Can he/she identify and label his or her feelings? Does the client experience mood fluctuation? At times counselors must clarify ambiguous feelings to ensure common understanding.

Cognition

Perhaps this section of the mental status evaluation is the most comprehensive because it focuses on the greatest number of variables. Note, however, that all areas of assessment in this section deal with the client's former or current thought processes. Here the counselor observes the client's alertness and responsiveness. Orientation in the three spheres (person, place, and time) is assessed. In other words, does the client know who he or she is, where he or she is, and what the date and time are? Is he or she thinking abstractly or concretely? For example, is the adult client's thinking regressed and compromised? Assess any delusions (false beliefs that may or may not appear to be realistic) or hallucinations (false perceptions that may be auditory, visual, tactile, or olfactory). Any disorganized thoughts? Are both recent and remote memory intact? Evaluate attention span. Get an impression of the client's level of intelligence. The client's reality testing can be understood via his or her belief system, the level of insight into his or her problem areas, and whether the client is able to exhibit good judgment. Lastly, question the client on any ideation (thoughts) or plans (intended behaviors) to harm self or others, and note any and all responses verbatim. If present, discuss "harm to self or others" issues immediately with your supervisor. Plan ahead with your supervisor regarding procedures for crisis or emergencies.

As you complete the mental status evaluation, reviewing the three areas of assessment will help solidify your diagnostic impression and treatment recommendations.

DIAGNOSTIC IMPRESSION

When you arrive at this point in the clinical interview, you are cognizant of the presenting problem, possess a somewhat detailed psychosocial history, and have a good idea of the client's current mental status. The information gathered in the intake interview provides supportive documentation for formulating a diagnostic impression.

Counselors have mixed feelings about diagnosing. After all, is it not labeling? And labeling theory (Becker, 1991) indicates that persons often live up—or down—to a label. Also, one never knows in whose computer banks a diagnosis ends up. Yet, with counseling included in the medical model, we often must diagnose in order to get paid. All insurers and the government require a diagnosis for reimbursement for services. Consider the great ethical and moral responsibility we have!

Licensure and certification laws in most states permit only independent practitioners (psychiatrists, psychologists, clinical counselors, and clinical social workers) to diagnose mental and emotional disorders without supervision. Your responsibility is to develop a diagnostic impression based on your extensive interview. Do not minimize the importance of your diagnostic impression.

Making a tentative diagnosis (an impression, not carved in stone) improves with practice, further training, and experience. Finally, diagnostic impressions are fluid based on the client's potential change in level of functioning.

The Diagnostic and Statistical Manual
of Mental Disorders

In the mental health field, the diagnostic reference is the *Diagnostic and Statistical Manual of Mental Disorders* (4th edition, text revision), published by the American Psychiatric Association (2000). We suggest that you view this nosology (classification system) as a "cookbook." Each recipe has certain ingredients, which, in certain combinations, produce a certain product—a diagnosis. Each product has a number attached to it for ease of reporting. Diagnoses are clustered by groupings that have some similarities in symptoms. Severity is determined by several factors, including intensity, duration, and type of symptoms.

Be thorough, responsible, and accurate in your diagnosis. Choose labels carefully because, as previously stated, labels are not always used to the client's advantage. Always consult with your site supervisor and other independent practitioner colleagues to verify your impressions.

You may find that your impressions change as you learn more about your client in subsequent sessions and as additional problems are discussed. It is your responsibility to note any modifications to the diagnosis in the client record.

The task of establishing a diagnostic impression must be taken seriously. No two individuals possess the same interpersonal, environmental, or genetic factors. For these reasons your ultimate decision should be individualized even though similarities may be present. We owe it to our clients to assess them in the present moment, with no preconceived notions that may fit a person to a diagnosis. This affords greater objectivity and fairness.

At this point, let's take time to briefly describe the *Diagnostic and Statistical Manual* (DSM-IV-TR) in more detail. As with previous editions, the fourth edition of the manual is compatible with the *International Classification of Diseases Manual* (ICD-10), published by the World Health Organization, which has as its goal the maintenance of a system of coding and terminology for all medical disorders, including those of a psychiatric nature (American Psychiatric Association, 2000). The DSM-IV-TR is the fourth in a series of nosologies published by the American Psychiatric Association (and is actually the fifth revision). Its biopsychosocial approach attempts to consider holistically and atheoretically all possible clinical syndromes and developmental and medical variables of patients and strives to facilitate our diagnostic and treatment planning (American Psychiatric Association, 2000). This is a tall order. And, certainly, the DSM-IV-TR operates from a medical model, which can be limiting to those of us in the mental health field, considering that there are other equally compelling models such as learning and developmental approaches to mental dysfunctions.

Five places or categories, called *axes*, are possible to address in the DSM-IV-TR. They include the following:

Axis I   Clinical Syndromes
             Other Conditions That May Be a Focus of Clinical Attention
Axis II   Personality Disorders
             Mental Retardation
Axis III   General Medical Conditions
Axis IV   Psychosocial and Environmental Problems
Axis V   Global Assessment of Functioning

These are fully described in the DSM-IV-TR as well as other texts. It is not our intent in this book to present a course on the DSM-IV-TR; rather, we wish merely to introduce you to the labeling system of the mental health field. We suggest that you review the books listed in the bibliography and reference sections of this chapter. The DSM-IV-TR may well become a constant companion and a key reference for many of you.

TREATMENT RECOMMENDATIONS

Treatment recommendations ideally are made in concert with the client. When we include the client in decision making, we begin the teaching aspect of therapy: that counseling is a mutual process in which clients have valued input and equal responsibility regarding their treatment outcomes. Much depends on diagnosis, level of client acuity, client age, issues of convenience, and the expertise, skills, and theoretical orientation of the therapist. Recommendations should be based on what services would facilitate client ability to cope effectively and efficiently with current and future demands and responsibilities. Again, because no two individuals are exactly alike, the decision regarding treatment recommendations is as individualized as the diagnosis.

The following is a treatment recommendation checklist to assist in choice of treatment recommendations:

1. Are the client's problems acute (intense) or chronic (constant over time)?

2. What limitations may impede services (such as finances, transportation, child care, employment, general health, time constraints, or the like)?

3. Does the seriousness of the client's situation demonstrate the need for a structured inpatient environment?

4. Can the client's concerns be handled on an outpatient basis?

5. Are there requirements from the legal system (parole board or probation department)? Or from the Department of Youth Services or Child Protective Services?

6. Are we providing the least restrictive treatment environment that allows the client to address his or her problems?

7. Would the client do better in individual, group, or family counseling (or a combination)?

8. Does the specific diagnosis indicate a greater success rate in a specific treatment modality?

9. Will the client likely need short- or long-term services?

10. Do you need to collaborate with other professionals (such as a psychiatrist, psychologist, or clergy)?

11. Will any restrictions on your time and schedule influence your treatment?

12. Are there presenting problems outside your scope of practice, necessitating transferring the client to another professional within your agency or to another specialized agency? What is the most timely and effective process?

13. Will managed care or an insurance company direct the type and frequency of services?

14. Is your agency equipped with the trained personnel and facilities to address the client's specific problem areas (such as a play therapy room or group therapy room)?

15. Are you able to provide any specific services the client has requested (such as bibliotherapy or hypnosis)?

16. What are the client's strengths and barriers in achieving therapeutic goals?

17. Must any special needs (such as sensory impairments or the need for an interpreter) be addressed?

We attempt to help clients achieve their optimal level of functioning. At times, deciding on treatment recommendations may involve some creativity on our part. Even though we aspire to meet their needs, this may not always be possible. Do the best you can with what resources you have available, always operating ethically. Be candid with clients as to what you can and cannot do. For example, if you discover that neither you nor your agency can address the client's presenting problem, you have the obligation to help that client find the necessary resources (ACA Code of Ethics, Section A.11.b). As always, consult with your site supervisor.

Appendix K provides a sample treatment plan form that follows a management by objective format, is as behaviorally specific as possible, and includes outcome objectives and results. Very often, it is reviewed and updated every three months. We have found that this sample treatment plan form falls nicely within the parameters of what managed care providers look for in treatment planning. Ideally, we formulate the treatment plan in conjunction with the client; thus, the treatment plan becomes a social contract between the counselor and client. Note that we provide signature spaces for both client and counselor (and supervisor). In the case of minors, we ask the minor to sign in addition to the parent. How often are children and adolescents offered the opportunity to sign something? Signatures are one way to enlist investment in therapy. We urge you to discuss your treatment plan with your supervisor.

Upon completion of the clinical interview, clients should feel hopeful and confident in your ability to help them achieve their goals in this, the first step in the treatment process.

CONCLUDING REMARKS

The intake is a dynamic process involving both counselor and client. Components of the process include client and problem description, psychosocial history, mental status evaluation, diagnostic impression, and treatment recommendations. The ideal is to include the client in every step as a unique and distinct partner in the therapeutic journey.

REFERENCES

AMERICAN PSYCHIATRIC ASSOCIATION. (2000). *Diagnostic and statistical manual of mental disorders* (4th ed.) (Text revision). Washington, DC: Author.

BECKER, H. (1991). *Outsiders: Studies in the sociology of deviance*. New York: Free Press.

BENJAMIN, A. (1987). *The helping interview with case illustrations*. Boston: Houghton Mifflin.

CORMIER, S., & CORMIER, B. (1998). *Interviewing strategies for helpers* (4th ed.). Pacific Grove, CA: Brooks/Cole.

COUNCIL FOR ACCREDITATION OF COUNSELING AND RELATED EDUCATIONAL PROGRAMS (CACREP). (2001). *The 2001 standards*. Available at: www.counseling.org/cacrep/2001standards700.htm

DEGOOD, D. E., CRAWFORD, A. L., & JONGSMA, A. E. (1999). *The behavioral medicine treatment planner*. New York: John Wiley & Sons.

DOUGHERTY, L. M., & CHAMBLIN, B. (1999). Assessment as an adjunct in psychotherapy. In P. A. Lichtenberg (Ed.), *Handbook of assessment in clinical gerontology* (pp. 91–110). New York: John Wiley & Sons.

FAIVER, C. (2001). Effective treatment planning. In E. R. Welfel & R. E. Ingersoll (Eds.), *The mental health desk reference* (pp. 83–87). New York: John Wiley & Sons.

FISHER, D., BEUTLER, L. E., & WILLIAMS, O. B. (1999). Making assessment relevant to treatment planning: The STS clinician rating form. *Journal of Clinical Psychology, 55,* 825–842.

FRANCES, A., & ROSS, R. (1996). *DSM-IV case studies: A clinical guide to differential diagnosis*. Washington, DC: American Psychiatric Association.

JENSEN, P. S. (1999). Fact versus fancy concerning the multimodal treatment study for attention deficit hyperactivity disorder. *Canadian Journal of Psychiatry, 44*(10), 975–980.

KADDEN, R. M., & SKERKER, P. M. (1999). Treatment decision making and goal setting. In B. S. McCrady & E. E. Epstein (Eds.), *Addictions: A comprehensive guidebook* (pp. 216–231). New York: Oxford University Press.

KING, R. A., & SCAHILL, L. (1999). The assessment and coordination of treatment of children and adolescents with OCD. *Child & Adolescent Psychiatric Clinics of North America, 8*(3), 577–597.

KLASSEN, A., MILLER, A., RAINA, P., LEE, S. K., & OLSEN, L. (1999). Attention-deficit hyperactivity disorder in children and youth: A quantitative systematic review of the efficacy of different management strategies. *Canadian Journal of Psychiatry, 44*(10), 1007–1016.

MARUISH, M. E. (Ed.). (1999). *The use of psychological testing for treatment planning and outcomes assessment* (2nd ed.). Mahwah, NJ: Lawrence Erlbaum Associates.

OTHMER, E., & OTHMER, S. C. (1994). *The clinical interview using DSM-IV, Vol. I: Fundamentals.* Washing-

ton, DC: American Psychiatric Press.

PALEG, K., & JONGSMA, A. E., Jr. (2000). *The group therapy treatment planner.* New York: John Wiley & Sons.

SELIGMAN, L. (1990). *Selecting effective treatments.* Alexandria, VA: American Counseling Association.

SPITZER, R. L., GIBBON, M., SKODOL, A. E., WILLIAMS, J. B., & FIRST, M. B. (Eds.). (1994). *DSM-IV casebook.* Washington, DC: American Psychiatric Press.

BIBLIOGRAPHY

ADLER, A. (1964). *Problems of neurosis.* New York: Harper & Row.

DAVIS, S., & MEIER, S. (2001). *The elements of managed care: A guide for helping professions.* Pacific Grove CA: Brooks/Cole.

EGAN, G. (1998). *The skilled helper. A problem-management approach to helping* (6th ed.). Pacific Grove, CA: Brooks/Cole.

EISENGART, S., EISENGART, S., FAIVER, C., & EISENGART, J. (1996). Respecting physical and psychosocial boundaries of the hospitalized patient: Some practical tips on patient management. *Rural Community Mental Health, 23*(2), 5–7.

EISENGART, S., & FAIVER, C. (1996). Intuition in mental health counseling. *Journal of Mental Health Counseling, 18*(1), 41–52.

FAIVER, C., INGERSOLL, E., O'BRIEN, E., & McNALLY, C. (2001). *Explorations in counseling and spirituality: Philosophical, practical, and personal reflections.* Pacific Grove, CA: Brooks/Cole.

JONGSMA, A. E., Jr., & PETERSON, L. M. (1995). *The complete psychotherapy treatment planner.* New York: John Wiley & Sons.

JONGSMA, A. E., Jr., PETERSON, L. M., & MCINNIS, W. P. (1996). *The child and adolescent psychotherapy treatment planner.* New York: John Wiley and Sons.

MOLINE, M., WILLIAMS, G., & AUSTIN, K. (1998). *Documenting psychotherapy: Essentials for mental health practitioners.* Thousand Oaks, CA: Sage.

OTHMER, E., & OTHMER, S. C. (1994). *The clinical interview using DSM-IV, Vol. II: The difficult patient.* Washington, DC: American Psychiatric Association.

ROESKE, N. (1972). *Examination of the personality.* Philadelphia: Lea & Febiger.

SOMMERS-FLANAGAN, J., & SOMMERS-FLANAGAN, R. (1993). *Foundations of therapeutic interviewing.* Boston: Allyn & Bacon.

TEYBER, E. (1997). *Interpersonal process in psychotherapy: A guide for clinical training* (3rd ed.). Pacific Grove, CA: Brooks/Cole.

ZIMMERMAN, M. (1994). *Interview guide for evaluating DSM-IV psychiatric disorders and the mental status examination.* East Greenwich, RI: Psych Products Press.

6

Psychological Testing

An Overview

Counselors often hesitate to include psychological testing in their assessment of clients; some misunderstand the benefits of testing, and many regard testing with some skepticism. Their attitudes may be attributed to the traditional exclusionary use of testing to select those who will be admitted or rejected, to decide who will pass or fail, or to separate people into categories with labels. The historical concept and use of testing as a limiting or exclusionary tool may appear to be antithetical to the counseling profession's current orientation, which includes respecting and appreciating individual and cultural differences, viewing the client holistically, encouraging growth toward each client's optimal level of functioning, and empowering the client to take responsibility for self. Certainly, too, psychological testing has come under fire from various groups that are rightfully concerned about test bias and misuse of test results in our multicultural society (Hood & Johnson, 2002).

We would like to suggest that the counseling process can integrate psychological tests by using them in an inclusionary and adjunctive, rather than exclusionary, approach. That is, testing may be used as an additional and important source of information within a comprehensive evaluation; this can assist both counselor and client in the following ways:

- Developing a more thorough understanding of the client
- Appreciating strengths, potentials, and limitations
- Identifying problem areas, including psychopathology

- Validating clinical impressions of our clients
- Making treatment plans and recommendations
- Setting goals and objectives for measurable treatment outcomes
- Evaluating client progress

We believe that if psychological testing is used sensitively and is respected as a component of the counseling process, it can be a therapeutic experience that helps to enhance the working alliance. In Chapter 5 we expressed a similar concept concerning the use of the initial intake assessment as an integral component of the counseling process, and as an opportunity to build the therapeutic alliance.

We view psychological testing as a sort of second opinion about what may be going on in the client's life at the moment, instead of as a final, irrefutable fact. Groth-Marnat (1999) reports that individuals may be referred for psychological testing due to treatment impasses or the desire to get an extra perspective. Likewise, testing may be used either to justify a level of treatment or to validate that an accurate diagnosis has been made and appropriate treatment has been provided to avoid potential litigation. As with any opinion, which may vary over time and may be influenced by physical, emotional, and/or situational factors, the results of testing should always be taken with a grain of salt. (You may find it helpful to review your statistics notes for the concepts of reliability and validity at this point. You may also want to refer to Section E of the ACA Code of Ethics and Standards of Practice.) During your counseling internship and throughout your career, you may have access to modern computer scoring systems and methods that provide instant, lengthy, and often excellent "diagnostic" printouts based on the results of psychological testing. The trick here is to use modern technology without depersonalizing the client in the process. As counselors, we must value our professional knowledge and skills, our personal intuition, and our clinical interpretations as our most important therapeutic tools. View the client as a person, not a piece of interesting data!

Administering and interpreting psychological tests require advanced training, and, in most settings, the responsibility for testing rests with licensed clinical or school psychologists or licensed clinical mental health counselors. As a counselor intern, your exposure to psychological testing will vary greatly, depending on your internship setting. At some agencies, interns interface only infrequently with clinicians and clients who are involved in psychological testing, whereas at others, interns are more actively engaged in the testing process. During your internship, a psychometrician or other professional who is competent in psychological testing should closely supervise your participation in this area. As a counselor intern in any setting, you should be familiar with the following basic areas of knowledge relevant to psychological testing:

- Human behavior, personality theory, physiological psychology, and psychopathology (in order to recognize a need for testing)

- Appraisal concepts, such as power versus speed, reliability, validity, standard deviation, and mean
- Types of tests available—intelligence, personality, aptitude, achievement, interest inventory, and organicity
- Specific data concerning the particular instrument selected, such as norms, standardization criteria, cultural sensitivity, and applicability for individuals under the Americans with Disabilities Act
- Ethics and procedures related to the administration and interpretation of psychological tests
- General interpersonal and environmental conditions and individual situations that may influence test results

In addition, we as counselors and counselor trainees need to keep current with technical developments in testing so that we will be able to select the most appropriate instrument to provide information for our clients about their individual problems and concerns (Anastasi, 1992). Anastasi (1992) recommends that counselors take workshops and continuing education courses and read professional journals and other literature to stay abreast of the rapid advances in psychological testing. *The Fourteenth Mental Measurements Yearbook* (Plake & Impara, 2001) and *Tests in Print V* (Murphy, Impara, & Plake, 1999) are valuable resources for counselors to familiarize themselves with current tests, their applicability to various client populations, and issues of reliability and validity.

We must also keep in mind that the testing experience itself affects the client and has the potential to be damaging as well as therapeutic. When clients are tested, results may be acutely influenced by clients' current interpersonal and environmental variables. Communication of test results is a particularly important aspect of testing, and we suggest that when you talk with clients about their tests, you try to:

- Help clients understand that the test is only one source of information and that the data gathered "suggests a hypothesis about the individual that will be confirmed or refuted as other facts are gathered" (Anastasi, 1988, p. 490).
- Relate test data to the client's functioning in real-life situations so that they are meaningful and helpful (Hood & Johnson, 2002).
- Emphasize that the test reflects the client's functioning in the past, but that this information may be used to make changes in the future (Hood & Johnson, 2002).

Many clients are anxious about tests and may need to be reassured concerning the purpose of the instrument and the information that will be elicited. As a counselor intern, you will need to attend carefully to your clients' feelings and thoughts about each aspect of the testing process. Taking a test evokes strong feelings and thoughts in most people, which provide fertile ground for exploration and the acquisition of new insight. One counselor intern told us that she had referred her client to the agency's consulting psychologist for an MMPI-2 to confirm a diagnostic impression of major depression and obsessive-compulsive disorder. The client was extremely upset afterward, saying, "That psychologist told me I am clinically depressed. How could he say that about me! I tried so hard on that test, too. I really thought I 'aced' it. I guess I can never do anything right." Another counselor intern learned never to test children during lunchtime, recess, or physical education class because their levels of motivation and interest may affect test results. An empathic response on the counselor's part helps to relieve anxiety, promote healing and growth, and build the therapeutic alliance. As an additional reference and solid foundation, we suggest counselors thoroughly review Section E of the American Counseling Association's *Code of Ethics and Standards of Practice* (1995).

Myriad psychological tests are extant (refer to Plake & Impara, 2001, *The Fourteenth Mental Measurements Yearbook*, for a complete listing with descriptions). Some will have little value to you during your counseling internship, whereas others will offer relevant information for planning the best course of action for your clients. Just as tests are designed to categorize people and their problems, behaviors, style, or preferences, psychological tests themselves can be categorized into three discrete groupings: (1) tests of intelligence, (2) tests of ability, and (3) tests of personality (Anastasi, 1988). We suggest that you take the time, either before your internship or early in the experience, to review these general categories of tests and their uses and purpose.

Boughner, Hayes, Bubenzer, and West (1994) surveyed members of the American Association for Marriage and Family Therapy and ascertained that they used 147 different standardized assessment instruments. The authors found that although respondents did not use standardized instruments often, the Myers–Briggs Type Indicator (MBTI) and Minnesota Multiphasic Personality Inventory-2 (MMPI-2) were most popular. Moreover, Bubenzer, Zimpfer, and Mahrle (1990) conducted a national survey of agency counselors to determine how often they used psychological tests. The following tests were used most frequently:

- Minnesota Multiphasic Personality Inventory (MMPI), now revised (MMPI-2)
- Strong Campbell Interest Inventory (SCII), now revised (Strong Interest Inventory [SII])
- Wechsler Adult Intelligence Scale-Revised (WAIS-R), now revised again (WAIS-III)
- Myers-Briggs Type Indicator (MBTI)
- Wechsler Intelligence Scale for Children-Revised (WISC-R), now in its third edition (WISC-III), soon to be fourth

Our own experience confirms that these tests appear to be the most frequently used in clinical settings. We now discuss these five tests and how you, as a counselor intern, may use them to help your clients.

THE MINNESOTA MULTIPHASIC PERSONALITY INVENTORY-2 (MMPI-2)

Hathaway and McKinley developed the original MMPI in 1943 as a tool to facilitate making differential diagnoses in a population of psychiatric inpatients (Anastasi, 1988). Although the instrument was found not to be helpful in classifying persons into distinct diagnostic categories, it nevertheless provided a wide range of data that allowed accurate conclusions about personality traits, emotional characteristics, and psychological problems (Graham, 2000). The MMPI was used extensively for several decades, generating more than 8,000 research studies (Graham, 2000; Hood & Johnson, 2002). However, 40 years after its publication, critics raised concerns about archaic terms and references, sexist language, and obsolete item content in the instrument (Graham, 2000). The test was revised in 1989 and an updated version, the MMPI-2, was published (Graham, 2000).

The MMPI-2 contains 567 questions that the respondent answers as "True," "False," or "Cannot Say." Interpretation of results is based on T-scores. Butcher and Williams (2000) report that T-scores of 65 or greater (unless otherwise noted) are significantly elevated, whereas T-scores of 60 to 64 are moderately elevated. Whiston (2000) cautions that although a T-score of 65 is considered the cutoff score, clinicians do not label an individual according to one elevated scale. Generally, the more elevated the T-scores, the more likely the test taker is to have psychological or psychiatric problems. MMPI-2 results are interpreted, however, according to their "profile," overall configuration, and "code-type," or combination of the two or three most highly elevated scales' scores. An enormous amount of in-depth information about an individual's personality, symptoms, behaviors, and proclivities can be provided through analysis and interpretation of these code-types.

MMPI-2 scores are arranged in 4 validity scales that assess the subject's attitude toward the test and 10 clinical scales that assess emotional states, social attitudes, physical condition, and moral values (Graham, 2000). Whiston (2000) notes the ongoing debate concerning racial bias on the MMPI-2 and Minnesota Multiphasic Personality Inventory-Adolescent (MMPI-A) and encourages examiners to be knowledgeable about ethnic differences on the validity and clinical scales to ensure proper interpretation. The validity scales include:

- L, or Lie, which measures responses attempting to create a favorable impression, but which are likely untrue ("faking good")
- F, or Validity, which indicates carelessness, eccentricity, malingering ("faking bad"), confusion, or scoring errors
- ?, or Cannot Say, which indicates the number of questions left unanswered; this scale tends to invalidate the entire test if its score is too high
- K, or Correction, which indicates "faking good" or defensiveness if high and "faking bad" or self-denigrating attitude if low

The K scale is used as a factor to adjust or correct other scale scores on the MMPI-2. Hood and Johnson (2002) note that in addition to the four central validity scales, three new validity scales have been developed that augment the interpretation of the test. They are TRIN (True Response Inconsistency, which measures acquiescence or negativity), VRIN (Variable Response Inconsistency, which measures random responding), and Back F (a "fake bad" measure for the back, or new, part of the inventory).

The 10 clinical scales include the following:

1. Hs: Hypochondriasis
2. D: Depression
3. Hy: Hysteria
4. Pd: Psychopathic deviate
5. Mf: Masculinity/femininity
6. Pa: Paranoia
7. Pt: Psychasthenia
8. Sc: Schizophrenia
9. Ma: Hypomania
10. Si: Social introversion

Numerous additional scales and subscales have been developed to augment and refine information obtained from MMPI-2 results; the most frequently used of these are generally included in computer-scored reports. The supplementary scales include the following:

- Anxiety (A) and Repression (R) Scales
- Ego Strength (Es) Scale
- MacAndrew Alcoholism-Revised Scale (MAC-R)
- Overcontrolled-Hostility (O-H) Scale
- Dominance (Do) Scale
- Social Responsibility (RE) Scale
- College Maladjustment (Mt) Scale
- Masculine Gender Role (Gm) and Feminine Gender Role (Gf) Scale
- Posttraumatic Stress Disorder (Pk) Scale
- Subtle-Obvious Subscales

Graham (2000) notes that examinees need answer only the first 370 items if just the standard scales are required for the assessment. Several researchers have also developed Critical Items Lists, which indicate areas warranting possible further evaluation by the mental health professional.

During your internship, the MMPI-2 may be useful to you in the following ways:

- You will have a source of information regarding your client's personality style or problems before your initial meeting with the client, so that you may modify your counseling interventions appropriately.

- You can use MMPI-2 results to confirm a diagnostic impression or to rule out a diagnosis. This may be especially helpful in determining whether to request a psychiatric consultation for psychotropic medication because some disorders respond better than others to psychopharmacologic intervention.

- The test is valuable for augmenting information on clients who are not very verbal or have difficulty articulating their thoughts and feelings.

- You can discuss the critical items with your client as a way to explore problem areas and to help the client disclose important thoughts and feelings.

Hood and Johnson (2002) suggest that counselors always discuss critical items with each client, especially those related to suicidal ideation. These authors write that clients may assume that the counselor is already aware of all their concerns because a question on the MMPI-2 alluded to an area, and may therefore not raise important issues during sessions.

As with most other personality tests, results may change over time based on clients' emotional level of functioning. You can therefore compare scores from tests taken at different times to assess your client's response to life changes, stressors, or previous treatment, and you can also request an MMPI-2 after counseling to evaluate progress.

THE STRONG INTEREST INVENTORY (SII)

The SII, formerly the Strong Campbell Interest Inventory (SCII), was revised in 1994. This inventory reflects clients' interests and capacities in the world of work and provides a dependable guide for career goals and development (Consulting Psychologists Press, 1994). The SII is the most researched interest inventory in counseling and is often considered a model for other interest inventories because of its psychometric characteristics (Whiston, 2000). It assesses individual likes, dislikes, and preferences for activities and also compares the similarities and differences of individual characteristics and preferences with those of a population already employed in a particular occupation. Many sources (Anastasi, 1988; Brown & Brooks, 1996; Hood & Johnson, 2002; Walsh & Betz, 2001) stress that interest inventories like the SII evaluate interests rather than abilities; therefore, they may be used only to predict how much an individual will enjoy an occupation or an academic area of study, rather than how likely he or she will be to succeed or how well he or she will do in a given field.

The SCII was first published in 1974 and represented a non–gender-biased revision that combined two earlier versions of this instrument, the Strong

Vocational Interest Blank for Men and the Strong Vocational Interest Blank for Women (Walsh & Betz, 2001).

Holland's theory, characterizing personality types and the type of work environment likely to be most congruent with each personality type, was the foundation for the structure of the SCII (Anastasi, 1988), and this continues to be the underlying SII basis. The instrument contains 317 questions, which yield data organized into the following categories:

- Four-color Profile with "snapshot" of results
- Six General Occupational Themes (RIASEC—Realistic, Investigative, Artistic, Social, Enterprising, and Conventional)
- 25 Basic Interest Scales
- Occupational Scales (109 occupations)
- Four Personal Style Scales
- Administrative Indexes

The Administrative Indexes provide information that may be used to help interpret the scale scores by addressing critical questions that must be answered to yield a confident interpretation, Infrequent Responses Scales indicating inappropriate response sets or respondent attitudes, Like/Indifferent/Dislike percentages, and the pattern of responses.

The General Occupational Themes scales record the client's overall preferences related to work environment, activities, coping style, and types of people the client likes to be with, and yield a three-letter Holland personality code-type, such as RIA or ASE (Walsh & Betz, 2001). These personality types described by Holland (1997) include the following:

- Realistic (oriented toward tools or machinery)
- Investigative (oriented toward scientific endeavors)
- Artistic (oriented toward self-expressive activities)
- Social (oriented toward helping other people)
- Enterprising (oriented toward goal-directed business)
- Conventional (oriented toward clerical activity)

The Basic Interest Scales report the client's interest in specific areas of activity, with higher scores indicating greater interest. The Occupational Scales reflect similarity of the client's interests to the likes and dislikes expressed by people in various occupations. A higher score indicates that the client has more in common with people in a particular occupation. The Personal Style Scales note particular environments in which individuals like to learn and work as well as the type of activities they find rewarding.

Hood and Johnson (2002) stress that clients who are self-motivated by a desire for information will benefit most from an assessment process using an interest inventory. These authors also caution that the SII and other similar instruments are not suitable for use with clients who have emotional problems;

these problems must be addressed before the client works on career or educational plans (Hood & Johnson, 2002). During your counseling internship, you may be able to use the SII to help your clients in the following ways:

- You can facilitate self-exploration and self-understanding as your client examines data regarding preferences, likes, and dislikes related to occupations, academic study, and other people.

- You can assist your client with educational plans related to career goals.

- You can provide new occupational options for your client to consider.

- You can help your client determine how much he or she has in common with other people who are employed in an occupational field.

- You can help validate your client's self-perceptions or clarify choices and thus build self-esteem.

THE WECHSLER ADULT INTELLIGENCE SCALE-III (WAIS-III)

The most widely used individually administered intelligence tests are the Wechsler instruments (Whiston, 2000). The Wechsler Adult Intelligence Scale, Third Edition (WAIS-III), published in 1997, represents a revision of the Wechsler Adult Intelligence Scale. The WAIS-III has the goal of establishing new norms within a more current representative sample, which includes expanded age ranges to reflect increased longevity (The Psychological Corporation, 1998). In addition, item content and item bias analyses were performed, artwork and materials were updated, and most subtests were altered specifically to assess individuals who have lower cognitive functioning. This instrument measures adult intelligence and is organized into a Verbal Scale, consisting of seven subtests, and a Performance Scale, consisting of seven subtests. These subtests are alternated, one from the verbal group and one from the performance group, as was the case with the WAIS-R.

The scores yielded are a Verbal Score, a Performance Score, and a Full Scale Score. The subtests are expressed as standard scores with a mean of 10 and a standard deviation of 3, whereas the Verbal, Performance, and Full Scale Scores are reported as deviation IQs with a mean score of 100 and a standard deviation of 15. The administration of the WAIS provides valuable anecdotal information about the examinee's task-approaching skills, level of concentration, and motivation (Whiston, 2000).

The WAIS-III Verbal Scale consists of the following subtests:

- Information (questions are related to general, basic information, arranged according to level of difficulty, from easiest to hardest)
- Digit Span (oral repetition of numbers, backward and forward)
- Vocabulary (words to be defined, arranged from easiest to hardest)

- Arithmetic (oral arithmetic problems)
- Comprehension (assesses understanding of meaning)
- Similarities (assesses ability to think in an abstract way)
- Letter-Number Sequencing (oral material that needs to be reordered and repeated): *supplementary subscale*

The WAIS-III Performance Scale consists of the following subtests:

- Picture Completion (cards in which the client must describe what is missing from each picture)
- Picture Arrangement (cards representing parts of a story that the client must arrange sequentially)
- Block Design (cards with colored patterns that the client must reproduce using colored blocks)
- Object Assembly (puzzles that must be assembled in a limited time): *optional subscale*
- Digit Symbol (a test of memory, eye-hand coordination, visual discrimination, and speed involving copying symbols for digits from a key card)
- Matrix Reasoning (examining geometric figures and either naming or pointing to the correct answer)
- Symbol Search (considering if target symbol groups appear within search symbol groups): *supplementary subscale*

The Wechsler Intelligence Scale may be used for differential diagnosis of psychological and mental problems, as well as for assessing intelligence (Anastasi, 1988; Hood & Johnson, 2002; Walsh & Betz, 2001). Variation among scores on subtests, differences among Verbal, Performance, and Full Scale Scores, and overall patterns of subtest and scale scores are clinically significant. Careful analysis of Wechsler Intelligence Scale test results by a highly trained, experienced clinician can provide information that may indicate the possibility of drug or alcohol abuse, brain damage, Alzheimer's disease, anxiety, depression, reading problems, antisocial behavior, and other psychological dysfunctions (Anastasi, 1988; Hood & Johnson, 2002).

During your internship you may use the WAIS-III to help your client in the following ways:

- You may be able to modify or structure your counseling style and interventions to meet your client's needs more appropriately if you are aware of his or her intellectual functioning and cognitive style. For example, you may opt to use a more behavioral, less insight-oriented approach with a client who is developmentally challenged.
- You may help a client in career or educational planning.
- You may be able to help a client access beneficial community resources, such as special public school or housing programs to which he or she is entitled, if you are aware of intellectual, neurological, or learning problems.

- You may be able to use the test scores to facilitate treatment planning by confirming a diagnostic impression or ruling out a possible diagnosis.
- You may be able to help clarify your client's problem. For example, is the source of the difficulty a cognitive, neurological, or emotional problem?

THE MYERS-BRIGGS TYPE INDICATOR (MBTI)

The most widely used personality instrument in the world today is the Myers-Briggs Type Indicator (Consulting Psychologists Press, 1998). Katherine Briggs and her daughter, Isabel Myers (Hood & Johnson, 2002), developed the Myers-Briggs Type Indicator in the 1920s, basing it on the work of Carl Jung. This instrument attempts to classify individuals into one of 16 specific personality types, using a four-letter code, and describes behavioral, emotional, and functional preferences. It contains 126 forced-choice questions. Hood and Johnson (2002) note that the MBTI is popular and widely used because there are no good or bad or higher or lower scores, and no dysfunctions or diagnostic categories are implied by the test results. Rather, the instrument indicates tendencies toward certain characteristics and preferences but allows that individuals may also be able to use the opposite characteristic or preference.

The MBTI assesses the following personality factors:

- Extroversion/introversion, an attitude describing the flow of psychic energy outward toward the world (E) or inward toward the self (I)
- Sensing/intuitive, a perceiving function of receiving information from the outside world in a sensing (S) or intuitive (N) way
- Thinking/feeling, a decision-making process based on data that have been received from the outside world in a thinking (T) or feeling (F) way
- Judging/perceiving, the individual's dominant means of relating to the outside world in a judging (J) or perceiving (P) way

An example of a Myers-Briggs code-type might be ENFJ, meaning that the individual is an extroverted person, with a preference for absorbing and processing information in an intuitive and feeling way, who generally relates to the world with a judging attitude. The MBTI four-letter code attempts to present a comprehensive portrait of the way these dimensions of personality balance one another in an individual. The instrument is relatively easy to score and interpret, and an instruction manual is included.

The MBTI was the first widely used personality instrument developed and scored using the Item Response Theory (Consulting Psychologists Press, 1998). The Item Response Theory (IRT) method yields the best estimate of a person's best-fit personality type, the level of performance of each test item, and the selection of test items having the most meaningful measurement.

Computer scoring methods also generate lengthy, detailed descriptions of client personality and behaviors. As always, we caution you to integrate test results as one data source within a comprehensive, evolving assessment of your client as a total human being. The MBTI is appropriate for use in a variety of counseling settings because it provides abundant data in a nonthreatening, interesting way and because advanced training is not necessary for the mental health professional who scores and interprets this instrument. As a counselor intern, you may use the MBTI to help your clients in the following ways:

- You may use knowledge of your own code-type as well as your client's code-type to modify your counseling interventions and improve counselor–client communication and the working alliance.

- You can help your client gain self-understanding and explore important issues by becoming aware of how his or her ways of perceiving and relating to the world affect thoughts, feelings, and behaviors.

- You can help couples and families improve interpersonal relationships by gaining an understanding of how each of their code-types and personality styles may affect communication and attitudes.

- You can help your client choose work environments that are congruent with his or her code-type.

- You can help promote health and growth in your client by exploring the less-used functions of his or her personality style and code-type as potential coping mechanisms, new solutions to problems, alternative ways of relating to others, and new concepts of self.

THE WECHSLER INTELLIGENCE SCALE
FOR CHILDREN-III (WISC-III)

The Wechsler Intelligence Scale for Children was originally developed as a direct downward extension of an adult intelligence assessment instrument, the Wechsler-Bellevue Intelligence Scale (Hood & Johnson, 2002). First revised in 1974 (WISC-R), the present 1991 (WISC-III) revision maintains the basic structure and content of the WISC-R. In addition, WISC-III provides updated normative data, improved design and items, and three supplementary subtests (The Psychological Corporation, 1997). This revision continues to attempt to maintain balanced sensitivity to multicultural and gender references.

The WISC-III is similar in structure, content, and scoring organization to the WAIS-III (discussed earlier in this chapter). This test, which assesses children ages 6 years through 16 years and 11 months, yields a Verbal Scale Score, a Performance Scale Score, and a Full Scale Score and contains 10 core and three supplementary subtests. The subtests are administered by alternating one from the Verbal Scale and one from the Performance Scale. The three supplementary subtests assess cognitive ability. The WISC-III subtests are also expressed as standard scores with a mean of 10 and a standard deviation of 3, whereas the Verbal, Performance, and Full Scale Scores are reported as deviation IQs with a mean of 100 and a standard deviation of 15.

The Verbal Scale of the WISC-III contains the following subtests:

- Information
- Similarities
- Arithmetic
- Vocabulary
- Comprehension
- Digit Span (supplementary subtest)

The Performance Scale subtests are:

- Picture Completion
- Picture Arrangement
- Block Design
- Object Assembly
- Coding
- Symbol Search (supplementary subtest)
- Mazes (supplementary subtest)

The WISC-III is scored, interpreted, and used in much the same way as the WAIS-III. This instrument can be used to assess children's intelligence, personality, and learning style, as well as for differential diagnosis of academic, neurological, and psychological problems. As a counselor intern, you may use the WISC-III to help your young clients in the following ways:

- You will be able to modify your counseling style and structure your interventions so that they will be more appropriate for your child client if you are aware of his or her cognitive and emotional developmental levels.
- You may be able to clarify your client's difficulties as cognitive, neurological, emotional, or a combination of factors.
- You may be able to facilitate treatment planning by confirming a diagnostic impression or ruling out a possible diagnosis.
- You may be able to help parents or caretakers access beneficial school or community-based programs for which the child is eligible.
- You may be able to help parents and school personnel understand and meet the child's individual needs more effectively.
- You may be able to assist parents, caretakers, and school personnel in developing a realistic picture of the child's potential so that they can set appropriate goals.

ADDITIONAL CLINICAL
ASSESSMENT INSTRUMENTS

Students interning at clinical settings that treat individuals with acute psychiatric or emotional problems, such as comprehensive mental health agencies and inpatient psychiatric facilities, need to be familiar with the following psychological tests. Remember, this brief summary is no substitution for reviewing your course materials, acquiring further knowledge about each specific test, and adhering to specific ethical and legal guidelines for the utilization of these tests. We believe the following psychological tests have merit, though what follows is not an all-inclusive list.

Rorschach Test

The Rorschach is the second most widely used projective testing technique and the second most widely used assessment instrument after the MMPI-2 (Cullari, 2001). Clients are shown 10 achromatic and chromatic inkblots and asked to describe their impressions of them. The examiner scores the responses according to characteristics or variables on each card such as form, color, shading, texture, movement, and content. Interpretations of the responses to these ambiguous inkblots help the counselor to understand the individual's personality. There are various scoring systems based upon the examiner's preference. Students should be aware that additional training and experience are required to become proficient in scoring the test.

Thematic Apperception Test

This widely used projective test consists of 31 possible cards dealing with various interpersonal situations (Murray, 1943). Individuals are presented with at least 20 of the cards and asked to make up a story about each picture and describe what is happening, what led up to this scene, what the individuals are thinking and feeling, and what the outcome of the story will be (Hood & Johnson, 2002). Examiners are able to look for themes in responses that may reflect interpersonal conflict, relationship issues, and family difficulties (Cullari, 2001). This test can be valuable for individuals who have difficulties accessing their feelings and are able to project their emotions onto the various cards.

Projective Drawings

Projective drawings offer an opportunity for the examiner to understand clients' feelings, thoughts, and behaviors by interpreting clients' art. There are various techniques, but most often clients are asked to draw a human figure; a house, a tree, and a person; or a drawing depicting their own family involved in an activity (Cullari, 2001; Halpern & McDay, 1998; Koppitz, 1984). Scores are assigned to various aspects of the drawings, including such factors as overall theme or feeling-tone of the picture, characteristics of the individuals, number of body parts depicted, erasures, placement on the page, and relationships between individuals or family members.

Sentence Completion

This assessment instrument involves having individuals complete sentence stems that express ideas, feelings, and concerns (Cullari, 2001). Ames & Riggio (1995) note that the results for sentence completions are viewed as reliable and valid with many populations. The Rotter Incomplete Sentences Blank (Rotter, Lah, & Rafferty, 1992) has formal administration and scoring systems. Examiners review responses with specific attention to certain ones, as well as to themes that might correlate with diagnostic or treatment issues. This assessment instrument is easy to learn and seems to be a simple assessment technique for most examiners. Individuals completing the tasks find it nonthreatening and a good "ice breaker" in counseling and testing situations.

ADDITIONAL CHILD AND ADOLESCENT ASSESSMENT INSTRUMENTS

There are various psychological assessment instruments that you may want to utilize when evaluating children and adolescents. Again, we encourage you to review your specific psychological testing coursework to complete your understanding of the ethical, legal, and testing standards applicable to these and all psychological assessments. The following tests are not inclusive of child and adolescent psychological instruments, but they are tests we find to be valuable.

Achenbach Child Behavior Checklist

Achenbach (1997) developed this instrument to measure behavior problems for children ages 4 to 18 years. The test yields scores in seven areas:

- Anxious/Depressed
- Withdrawn
- Somatic Complaints
- Social Problems
- Thought Problems
- Attention Problems
- Aggressive and Delinquency

There are separate rating scales for a parent/guardian report, teacher report, self-report, direct observation, and interview (Whiston, 2000). This instrument is an extremely valuable and popular behavior checklist and is used in both clinical settings and schools (Cullari, 2001; Hood & Johnson, 2002).

Million Adolescent Clinical Inventory (MACI)

This clinical inventory is the "junior version" of the Million Clinical Multi-axial Inventory (MCMI), which is used for adults as an alternative to the MMPI-2 for diagnosing psychopathology (Million & Davis, 1993; Million, Million, & Davis, 1994). These inventories provide scores that correspond with Axis I (clinical syndromes) and Axis II (personality disorders) on the DSM-IV. Hood and Johnson (2002) advise students that both inventories require a considerable amount of training and experience to utilize them properly.

Children's Depression Inventory (CDI)

This easy to administer, score, and interpret test is one of the most highly researched instruments designed to assess depression in children (Hood & Johnson, 2002). It is a self-report inventory measuring depression in children and adolescents ages 8 to 17 years (Kovacs, 1992). There are 27 self-report items that are similar to the Beck Depression Inventory (BDI) items, assessing Negative Mood, Interpersonal Problems, Ineffectiveness, Anhedonia, and Negative Self-Esteem.

Conners' Rating Scales-Revised

Conners (1997) developed rating scales to evaluate behavioral and emotional problems and attention deficit/hyperactivity (ADHD) in children and adolescents ages 3 to 17 years. Each scale assesses the appropriateness or inappropriateness of the following: anger control, oppositional behavior, conduct, cognitive/inattention problems, emotional problems, hyperactivity, anxious/shy, family problems, perfectionism, social problems, psychosomatic concerns, and ADHD (DSM-IV). Scores for the scales are based on parents' and teachers' observations of the child's behavior. There is also a self-assessment version for adolescents. School psychologists frequently use the Conners' scales for identifying developmental disorders in students (Zaske, Hedstrom, & Smith, 1999). In addition, the scales are used for routine screening in mental health agencies, schools, pediatricians' practices, and many other settings.

Minnesota Multiphasic Personality Inventory-Adolescent (MMPI-A)

This test is different from the MMPI-2, based upon norms, item content, and nature of several of the scales (Hood & Johnson, 2002). The test targets adolescents 14 to 18 years of age and provides separate same-sex norms. To encourage adolescent cooperation there are fewer items than in the MMPI-2. There are similarities between the MMPI-A and MMPI-2 in 11 content areas, but four scales (School Problems, Low Aspirations, Alienation, and Conduct Disorder) address specific adolescent issues (Hood & Johnson, 2002).

Wechsler Preschool and Primary Scale
of Intelligence-Revised (WPPSI-R)

This intelligence instrument assesses children 4 to 6½ years of age (Hood & Johnson, 2002). The test was revised in 1989 from the Wechsler Preschool and Primary Scale of Intelligence (WWPSI) (Wechsler, 1989). There are 11 subtests with 10 used to generate an intelligence quotient. Eight subtests are similar to those of the WISC. This instrument has the same normalized standard scores for subtests and intelligence quotients as the other Wechsler tests.

CONCLUDING REMARKS

Psychological testing can serve as one valuable source of data and an important component of a comprehensive assessment of the client. The testing experience can promote self-understanding and growth and can serve to strengthen the therapeutic alliance if you, as the counselor intern, attend to your client's feelings, thoughts, and behaviors throughout the testing process. Psychological testing should be viewed as a second opinion, a means of augmenting information and of increasing your understanding of the client's current concerns and current level of functioning. Results should be taken with a grain of salt! We believe that test results should be used to provide more pieces to the puzzle in understanding your client and should never be used as a means of disposition for any client.

REFERENCES

ACHENBACH, T. M. (1997). *Child behavioral checklist for ages 4–18*. Itasca, IL: Riverside Publishing.

AMES, P. C., & RIGGIO, R. E. (1995). Use of the Rotter Incomplete Sentence Blank with adolescent populations: Implications for determining maladjustment. *Journal of Personality Assessment, 64,* 159–167.

ANASTASI, A. (1988). *Psychological testing* (6th ed.). New York: Macmillan.

ANASTASI, A. (1992). What counselors should know about the use and interpretation of psychological tests. *Journal of Counseling and Development, 70,* 610–615.

BOUGHNER, S., HAYES, S., BUBENZER, D., & WEST, J. (1994). Use of standardized assessment instruments by marital and family therapists: A survey. *Journal of Marital and Family Therapy, 20,* 69–75.

BROWN, D., & BROOKS, L. (Eds.). (1996). *Career choice and development* (3rd ed.). San Francisco: Jossey-Bass.

BUBENZER, D., ZIMPFER, D., & MAHRLE, C. (1990). Standardized individual appraisal in agency and private practice: A survey. *Journal of Mental Health Counseling, 12,* 51–66.

BUTCHER, J. N., & WILLIAMS, C. L. (2000). *Essentials of MMPI-2 and MMPI-A interpretation* (2nd ed.). Minneapolis: University of Minnesota Press.

CONNERS, C. K. (1997). *Conners' rating scales-revised*. North Tonawanda, NY: Multi-Systems, Inc.

CONSULTING PSYCHOLOGISTS PRESS, INC. (1994). *Strong Interest Inventory application and technical guide.* Palo Alto, CA: Consulting Psychologists Press, Inc.

CONSULTING PSYCHOLOGISTS PRESS, INC. (1998). *1998 preview: Myers-Briggs Type Indicator revision.* Palo Alto, CA: Consulting Psychologists Press, Inc.

CULLARI, S. (2001). Counseling and psychotherapy: A practical guidebook for students, trainees and new professionals. Needham Heights, MA: Allyn & Bacon.

GRAHAM, J. (2000). *MMPI-2: Assessing personality and psychopathology* (3rd ed.). New York: Oxford University Press.

GROTH-MARNAT, G. (1999). Financial efficacy of clinical assessment: Rational guidelines and issues for future research. *Journal of Clinical Psychology, 55*, 813–824.

HALPERN, J., & MCKAY, K. (1998). Psychological testing for child and adolescent psychiatrists: A review of the past 10 years. *Journal of the American Academy of Child and Adolescent Psychiatry, 37*, 575–584.

HOLLAND, J. (1997). *Making vocational choices: A theory of vocational personalities and work environments* (3rd ed.). Odessa, FL: Psychological Assessment Resources.

HOOD, A., & JOHNSON, R. (2002). *Assessment in counseling: A guide to the use of psychological assessment procedures* (3rd ed.). Alexandria, VA: American Counseling Association.

KOPPITZ, E. M. (1984). *Psychological evaluation of human figures drawings by middle school pupils.* Orlando, FL: Grune & Stratton.

KOVACS, M. (1992). *Children's Depression Inventory manual.* North Tonawanda, NY: Mental Health Systems.

MILLION, T., & DAVIS, R. D. (1993). The Million Adolescent Personality Inventory and the Million Adolescent Clinical Inventory. *Journal of Counseling and Development, 71*, 570–574.

MILLION, T., MILLION, C., & DAVIS, R. D. (1994). *Million Clinical Multiaxial Inventory-III: Manual for MCMI-III.* Minneapolis: National Computer Systems.

MURPHY, L. L., IMPARA, J. C., & PLAKE, B. S. (Eds.). (1999). *Tests in print V* (Vol. II). Lincoln, NE: The Buros Institute of Mental Measurements, The University of Nebraska.

MURRAY, H. A. (1943). *Thematic Apperception Test.* Cambridge, MA: Harvard University Press.

PLAKE, B. S. & IMPARA, J. C. (Eds.). (2001). *The fourteenth mental measurements yearbook.* Lincoln, NE: The Buros Institute of Mental Measurements, The University of Nebraska.

THE PSYCHOLOGICAL CORPORATION. (1997). *Wechsler Intelligence Scale for Children* (3rd ed.). San Antonio, TX: The Psychological Corporation.

THE PSYCHOLOGICAL CORPORATION. (1998). *Wechsler Adult Intelligence Scale* (3rd ed.). San Antonio, TX: The Psychological Corporation.

ROTTER, J., LAH, M., & RAFFERTY, J. (1992). *Rotter Incomplete Sentences Blank: Manual* (2nd ed.). San Antonio, TX: The Psychological Corporation.

WALSH, W., & BETZ, N. (2001). *Tests and assessments* (4th ed.). Upper Saddle River, NJ: Prentice-Hall.

WECHSLER, D. (1989). *Manual: Wechsler Preschool and Primary Scale for Intelligence.* San Antonio, TX: The Psychological Corporation.

WHISTON, S. (2000). *Principles and applications of assessment in counseling.* Belmont, CA: Wadsworth.

ZASKE, K. K., HEDSTROM, K. J., & SMITH, D. K. (1999, August). *Survey of test usage among clinical and school psychologists.* Paper presented at the 107th Annual Convention of the American Psychological Association, Boston.

BIBLIOGRAPHY

COOPER, S. (1995). *The clinical use and interpretation of the Wechsler Intelligence Scale for Children* (3rd ed.). San Antonio, TX: The Psychological Corporation.

GROTH-MARNAT, G. (1997). *Handbook of psychological assessment* (3rd ed.). New York: Wiley.

HANSEN, J. C. (2000). Interpretation of the Strong Interest Inventory. In C. E. Watkins, Jr., & V. L. Campbell (Eds.), *Testing and assessment in counseling practice* (2nd ed., pp. 227–262). Mahwah, NJ: Erlbaum.

KAUFMAN, A. (1994). *Intelligence testing with the WISC-III*. San Antonio, TX: The Psychological Corporation.

KRONENBERGER, W. G., & MEYER, R. G. (2001). *The child's clinician's handbook* (2nd ed.). Boston: Allyn & Bacon.

NEWMARK, C. (1996). *Major psychological assessment instruments* (2nd ed.). Boston: Allyn & Bacon.

THE PSYCHOLOGICAL CORPORATION. (1997). *WAIS-III-WMS-III technical manual*. San Antonio, TX: The Psychological Corporation.

WALLACE, S., & LEWIS, M. (1998). *Becoming a professional counselor* (2nd ed.). Thousand Oaks, CA: Sage.

WECHSLER, D. (1997). *The Wechsler Adult Intelligence Scale: Administration and scoring manual* (3rd ed.). San Antonio, TX: The Psychological Corporation.

7

Understanding How
to Help

By the time you are ready to begin your internship, you will have completed all or most of the academic courses required for your degree in counseling. You will have studied a variety of theoretical models and clinical techniques, and you will have acquired a substantial knowledge base in other academic areas that are relevant to counseling. However, as a counselor intern, you may be wondering how to begin helping the client who presents with complicated problems and says, "I'm a hopeless case. I've already tried everything and nothing has worked." In this chapter, we provide a practical framework to help you synthesize the academic knowledge you have acquired and formulate a rational protocol for helping clients. We include discussions of the counseling process, the therapeutic relationship, theoretical orientation, specific counseling techniques we feel have the most salience for you as a counselor intern, and useful treatment modalities.

THE COUNSELING PROCESS

Counseling may be conceptualized as an interactive process involving a trained professional counselor (or counselor intern) and a client, with the purpose of enhancing the client's level of functioning. Counselor and client work together in a collaborative fashion, helping the client grow and change by identifying goals, developing new ways of understanding and coping with problems, and learning to use internal and environmental resources more effectively.

When beginning to work with clients during your internship, it is helpful to remember that the counseling process always involves the following stages, regardless of your theoretical orientation, your level of clinical experience, or the complexity of the client's problems:

- Relationship-building (developing trust and building a therapeutic alliance)

- Exploration (developing a full understanding of client and problems)

- Decision-making (deciding on specific goals and strategies to achieve them)

- Implementation (utilizing internal and external resources to resolve concerns)

- Termination and follow-up (assessing readiness to end and ensuring that progress is maintained) (Doyle, 1998)

These stages are interrelated and interlocking, rather than linear or sequential. Clients may need to move through some of them more than one time. For example, difficulties with implementation may lead to further work on decision-making or exploration.

THE THERAPEUTIC RELATIONSHIP

The relationship between counselor and client is of paramount importance in determining the effectiveness of counseling (Corey, 2001b). It is the key therapeutic ingredient, as well as the primary means of inspiring positive change in clients (Corey, 2001b; Holmes, 1999; Kahn, 1997). Effective counseling is more than a theoretical approach or a particular set of counseling skills. A high-quality therapeutic relationship is healing for the client, regardless of the counselor's theoretical orientation or use of specific counseling techniques (Corey, 2001b).

Counselor empathy has been recognized as the most central dynamic in determining the quality of the therapeutic relationship (Kahn, 1997; Shea, 1998). Attending carefully to the client's thoughts, feelings, and behaviors; respecting the client's worth and dignity; and using *active listening* skills are important facets of counselor empathy and provide means to foster a high-quality therapeutic relationship. Active listening will help you to better understand the client's subjective experience and will demonstrate to the client that you are trying hard to understand his or her concerns. During your internship, your ability to be empathic and to convey your care, concern, and respect will be healing for your client.

Counselor trustworthiness is another healing force in the therapeutic relationship. When your client trusts you, he or she will feel safe enough to explore important feelings, develop new perspectives, and try out new behaviors. Prove your trustworthiness in basic, concrete ways. Always be reliable and dependable. We have found the following behaviors to be especially helpful in developing client trust:

- Make certain that your client has your full attention and respect during sessions, and try to demonstrate your caring in both verbal and nonverbal ways.

- Start and end sessions on time, and do your best to maintain the regularity of the appointment schedule that you and your client have agreed upon. Whenever possible, give your client plenty of advance notice if you will be on vacation or otherwise unavailable for regular sessions.

- Always follow through on counseling-related tasks that you have discussed with your client, such as finding a support group or a list of helpful books.

- Maintain strict confidentiality. Never discuss a client outside of the treatment setting unless you first have a signed release and have explained to your client exactly what information will be disclosed or exchanged.

- Do not talk about the specific details of problems of other clients, even without mentioning their names. Clients may feel that their own time with the counselor is intruded upon when attention is focused on other clients during the session. Clients also may worry about a breach of confidentiality if the counselor discloses certain aspects of other clients' problems.

- Maintain professional boundaries at all times. It is unwise to engage in social relationships or participate in any business transactions with clients outside of the counseling setting (Corey, Corey, & Callahan, 1998). Do not disclose intimate personal information or discuss your own problems, as clients may feel obligated to take care of your needs and emotions, rather than attending to their own (Corey, Corey, & Callahan, 1998).

YOUR THEORETICAL ORIENTATION

Beginning counselors are not always aware that the theoretical orientation one selects provides the lens through which one views the client and the framework for conceptualizing the client's issues, as well as the structure for formulating interventions. In the sections that follow, we discuss several important theoretical schools and some of the counseling techniques and interventions that are associated with each one. However, it is important to remember that the theoretical approach you choose is much more than a set of specific counseling techniques or interventions; it is a schema for your way of being with clients, as well as for your way of understanding them.

As a counselor intern, you will be working to develop a repertoire of therapeutic techniques and skills to help your client. Although a particular counseling technique may be specific to one or another of the theoretical schools, an integrative perspective is often very helpful for the student counselor (Corey, 2001a). A wide range of therapeutic techniques and flexibility in selecting them will provide you with diverse methods to assist clients during your internship. The interventions you choose should feel comfortable to you and should be appropriate for the particular client, the client's unique concerns, and the current stage of the counseling process. Your counseling techniques should be grounded in your theoretical orientation and should be guided by:

- A conceptualization of the client's issues
- An assessment of available resources and support
- The client's motivation to change
- The present stage of the counseling process
- Your evolving relationship with the client
- Your personal philosophy of life and your own understanding of humankind

Psychodynamic Therapies

Many important concepts and views introduced by psychoanalytic theorists form the underpinnings for other models of therapy, as well as for our current understanding of human behavior. However, classic psychoanalytic therapy requires a higher level of training for you and a larger investment of time and money for your client than is realistically possible during your internship. Moreover, due to its long-term nature and the frequency of provider contact required, psychoanalysis is not authorized for reimbursement by most agencies and third-party payers. In the past decade, a short-term approach known as *brief psychodynamic therapy* has been introduced, which is more congruent with the mental health care system in the United States and also has a number of outcome studies supporting its effectiveness (Messer, 2001). We feel that the psychodynamic concepts described in this section will prove very useful to you as a beginning counselor, no matter what your theoretical orientation.

Corey (2001a) writes that psychodynamic thinking examines the past for insight into current problems. As a counselor intern, you can appraise your client's developmental history to gain a more thorough understanding of his or her present level of functioning. Taking the psychosocial history is an important part of your client assessment and is discussed in more detail in Chapter 5, The Clinical Interview.

Psychodynamic concepts of *transference* and *countertransference* provide useful avenues for understanding and helping your clients during your internship. In counseling, transference is a process whereby the client unconsciously transfers certain aspects of past or present interpersonal relationships onto the current relationship with the counselor. Transference refers to the client's unconscious tendency to experience feelings, attitudes, longings, and fears toward the counselor that were originally felt for other important people in the client's life (Kahn, 1997). For example, a woman who was raised in a family where feelings of anger were not accepted may begin to feel and express anger toward her counselor during her counseling sessions. Transference also includes the client's unconscious tendency to attribute attitudes to the counselor that in reality are, or were, held by other important people in the client's life. For example, the client whose wife is highly critical of him may begin to perceive the counselor as critical, when in fact the counselor has no feelings of disapproval for the client.

Countertransference refers to all the reactions or feelings the counselor experiences toward the client, which may be due to the client's behavior or to the counselor's own issues (Kahn, 1997). Countertransference is useful to you as a counselor intern, because it may help you understand how your client's behavior affects other people, and how in turn other people's reactions affect your client. For example, if you frequently find yourself feeling irritated by a client, you may understand more clearly why this client complains that people are always avoiding him or her. On the other hand, you may be feeling irritated because the client reminds you of someone from your own past. Self-knowledge and awareness of your own issues are essential, so that you can monitor your countertransference feelings and be certain that they do not interfere with your effectiveness as a counselor.

Although psychodynamic therapists use transference extensively as a therapeutic means to allow clients to "work through" their problems, we suggest that simply understanding the concept of transference may allow you to be more helpful to your client during your internship. For example, if your client expresses hurt because he or she perceives you as uncaring during a session, when you in fact have warm, caring feelings for the client, you should first examine your behavior to be certain you have not really done or said anything to give this client the feeling that you are uncaring. Next, you should empathize and express to the client how uncomfortable it must be to perceive the counselor as uncaring. Finally, you should explore the client's feelings by asking about your counseling relationship in the present and also by inquiring whether the client has ever felt that other people have treated him or her in an uncaring way (Kahn, 1997).

The psychodynamic term *object relations* refers to "interpersonal relations as they are represented intrapsychically" (Corey, 2001c, p. 83). Freud used the word *object* to refer to the significant person or thing that is the object of our feelings, wishes, or needs. Object relations of early life are replayed throughout the life cycle, even in adult interpersonal relationships, because people unconsciously seek reconnection with their parents and therefore repeat early childhood patterns of interaction (Corey, 2001c). In other words, we all tend to repeat, to some extent, the interpersonal dynamics of our family of origin. You can use this psychodynamic insight to encourage your client to become aware of repetitive patterns in relationships. Often, a client can be helped to change a maladaptive behavior by understanding where the behavior is originating. For example, a man who chooses emotionally distant, critical partners over and over again can be helped to understand that he may be re-creating his own early, painful relationship with an unavailable or rejecting parent. With this knowledge, you and your client can work on issues such as self-esteem, anger, shame, loss, and improved coping skills, so that he does not need to go on choosing damaging relationships.

A familiarity with the defense mechanisms described in the psychodynamic literature will serve you well as you work to understand clients. We suggest that you take some time to review these and remember also that your client will use a defense mechanism unconsciously and automatically for protection when he or she is feeling overwhelmed, threatened, or otherwise unsafe (Gabbard, 2000). Your client's use of defense mechanisms is a signal for you to take notice. Perhaps the material is too difficult and you need to slow down or wait before addressing it. On the other hand, perhaps the client is avoiding the reality of painful issues and needs to be confronted or encouraged to explore the material.

Heinz Kohut (1977), a psychodynamic theorist, writes that human beings require that *three basic needs* be met in order to develop a complete sense of self. These are:

- The need for *mirroring*, which means being valued and approved of
- The need for an *idealized other*, which means having another person to turn to for comfort and safety
- The need for *belonging*, which means feeling like other people and being accepted as part of a group

Kohut (1977) explains that we have these needs throughout our lifetimes and that often our sense of well-being is determined by our ability to find other people, whom he terms self-objects, who can meet our needs. Although ideally these needs are met by parents during early childhood, an incomplete sense of self can be repaired even in adulthood if our needs are met by a caring other, such as a therapist (Kohut, 1977).

Kohut's ideas, which form the basis for his *self psychology*, have significance for you as a counselor intern in three ways. First, recognizing the underlying unmet needs that are causing your clients pain helps you to be more empathic. Second, you may understand more clearly what sort of therapeutic response would be most healing for each client in light of his or her unmet needs. This allows you to formulate your interventions more effectively. Third, you can provide hope and a sense of empowerment to your clients by helping them understand that their needs and yearnings are universal and that there is a way to feel better by learning to establish healthy relationships throughout their lives.

Cognitive-Behavioral Therapies

Cognitive-behavioral therapies are applicable in many diverse counseling settings, and the underlying concepts and techniques are well-suited for you to use during your internship. Theorists such as Ellis, Beck, and Meichenbaum maintain that our thoughts influence our emotions and behaviors, so that the way we think determines how we feel and how we act (Corey, 2001a, 2001b). Cognitive-behavioral teachings highlight the role of positive, rational thinking in maintaining a sense of well-being and good mental health. In cognitive-behavioral therapy, the clients' dysfunctional thought patterns and maladaptive conclusions about themselves and their environments are examined, tested against contradictory evidence, challenged, and replaced with more adaptive beliefs, leading to therapeutic change (Beck & Weishaar, 2000). As a counselor intern, you will be able to empower your clients by demonstrating to them that they can help themselves feel better and make positive life changes by changing the way they think.

Albert Ellis (2000) describes his A–B–C theory, in which he identified irrational beliefs, rather than actual circumstances or events, as the causative factors in how we feel and act. He writes that A represents the *activating* event, B represents out *beliefs* about the event, and C represents the *consequences* evoked by our beliefs. Typically, these consequences are feelings and actions. This theoretical concept may be stated more simply as: "It depends on how you look at things" or "You can picture the glass as either half empty or half full."

During your internship, you may be able to use a simple story to illustrate the A–B–C link between thinking and feeling for your clients. For example, we ask our clients to imagine being awakened at 5 A.M. by the sound of someone opening the front door. We ask them how they would feel if they thought an intruder had broken into their home, and most clients answer that they would feel afraid and anxious. Next, we ask them how they would feel if they were awakened by the same sound at 5 A.M., but remembered that a family member had planned to start out extra early for work that day. Most clients respond that they might feel calm or relieved at that thought, but would not feel afraid or anxious. We discuss with our clients the difference in feelings evoked by different thoughts about the very same situation (being awakened at 5 A.M.) and the very same sound (someone opening the front door).

As a counselor intern, you can use cognitive restructuring with your clients by helping them identify their basic irrational beliefs and automatic negative thoughts. You can help them to "think things through" in a more rational, positive manner. Often, you will help them explore alternative attributions for other people's behaviors, as well. Cognitive-behavioral theorists advocate using logical questions and a collaborative stance, so that you and the client are a team, working together. For example, perhaps a client reports feeling depressed because a friend has not telephoned all week. The client assumes that the friend no longer likes him or her, and then concludes, "I must be unattractive and unlikable." As a counselor intern, you may ask your client to try to imagine other explanations for the friend's delay in calling. Could this friend have had a busy week? Been out of town? Tried to call but got a busy signal or missed finding the client at home? Could it be that the friend simply forgot to call?

You may ask your client what evidence there is that the friend no longer likes him or her, even if the friend has not called all week. When the client begins to explore other reasons for the friend's behavior and begins to think of the possibility that the friend may in fact like the client despite not calling, the client will begin to feel less depressed. You can go on using a cognitive-behavioral approach by questioning whether the friend's opinion of the client actually would alter the client's self-worth. After all, the client would remain the same, regardless of this particular friend's opinion. You can then help the client challenge his or her belief that self-worth depends solely on another person's affection. The client's maladaptive thoughts led to the depressed feelings; restructuring those thoughts leads to an improved mood. With practice, your client can learn how to think things through even outside of your counseling sessions. Cognitive-behavioral therapists call this technique *self-talk*.

Another effective cognitive-behavioral technique is called *catastrophizing*. In practical terms, that means asking your client to verbalize the worst possible scenario and then to explore consequences and coping mechanisms. This is another way of teaching clients to think things through in a rational, logical manner, rather than allowing irrational thoughts to stir up feelings of distress. For example, one of our clients, a 20-year-old student, told us she felt afraid to join in class discussions even when she knew the material well and was very interested in the subject matter. She related her fear to worry that she would give the wrong answer. Closely connected to this thought was the assumption that her instructor and classmates would disapprove if her opinion were different from theirs. The student's irrational belief was: "I must not participate in class discussions unless I am sure my opinion is the same as that of the other people in my class, or I will make a fool of myself." After empathizing with this client's fears, we asked her to imagine what would really happen if she gave her own opinion during a class discussion. She acknowledged that being informed and interested would likely lend validity to her opinion. She decided that even if the instructor and the other students had different opinions, they would not judge her to be stupid.

One possible pitfall of cognitive-behavioral therapies is that they may neglect exploration of emotional issues and may lead to too much intellectualization on the part of both counselors and clients (Corey, 2001a). Corey (2001a) advises that this type of therapy is not suitable for clients of limited intellectual ability. In addition, he warns that counselors may tend to impose their own values and views on clients or may come across as trying to invalidate the client's feelings when using cognitive-behavioral techniques. Clients may misinterpret the counselor's efforts to challenge their irrational beliefs and may see the counselor in a punitive, disapproving, or parental role. As a counselor intern, you need to take great care to be empathic, to maintain your therapeutic relationship, and to emphasize a collaborative approach when you use cognitive-behavioral treatment modalities.

Humanistic/Existential Therapies

Humanistic/existential schools emphasize a phenomenological, experiential approach to counseling. These therapies, which draw on spiritual beliefs and existential philosophies, as well as psychological principles, stress the quality of the therapeutic relationship itself, rather than a set of therapeutic techniques, as an avenue toward client self-awareness and change (Corey, 2001a). The counselor's abilities to grasp the client's subjective internal experience and to relate to the client in an authentic manner are considered to be of prime importance (Shea, 1998).

Person–Centered Therapy Carl Rogers suggests that all human beings have an actualizing tendency toward growth, health, and the fulfillment of potential, and his *person-centered therapy* proposes that this tendency can be nurtured by the provision of a therapeutic relationship characterized by genuineness, unconditional positive regard, and empathy (Raskin & Rogers, 2000). Rogers' teachings have two very significant implications for you during your internship. First, Rogers' belief in an inherent, universal human potential for self–actualization means that, as an intern, you do not need to be too concerned about diagnostic categories or labels in order to decide how to build a relationship with clients. You will be able to approach most clients in the same way. Second, you will be assured that even though you have relatively little clinical experience or counseling practice, you will be able to help your clients by interacting with them in an authentic, accepting, and empathic manner.

Genuineness, Rogers' first characteristic of the optimal therapeutic relationship, means being in touch with your own inner experiences and feelings and presenting yourself as a real person, rather than retreating behind a professional facade. Being genuine, or congruent, does not mean that you should reveal all aspects of your personality or information regarding your private life to your client, nor does it mean that you should express every thought and feeling that you experience during the session. Instead, being genuine, for the clinician, means being responsive, being appropriately spontaneous, being consistent in relating to the client, and not hiding behind a professional persona (Raskin & Rogers, 2000; Shea,1998).

The second aspect of Rogers' therapeutic relationship is *unconditional positive regard*. This concept refers to the counselor's acceptance of all the client's feelings and thoughts in a caring and nonjudgmental way; the counselor must respect and prize the client (Raskin & Rogers, 2000). As an intern, you sometimes will counsel clients whose attitudes, values, feelings, and behaviors conflict with your own beliefs and ideals. Your goal and responsibility in these situations is to maintain your unconditional positive regard for the client as a unique and valuable human being.

Empathy is the third relational quality that Rogers found to be essential in counseling. Empathy may be conceptualized as the counselor's ability to be emotionally attuned to the client, and the counselor's ability to recognize, understand, and share the client's internal world. Many authors (Kahn, 1997; Raskin & Rogers, 2000; Shea, 1998) point out that empathic healing is actually an interpersonal, interactive process. The counselor must not only be empathic, but also be able to convey that empathic caring and interest, so that the client is able to perceive it clearly. As a counselor intern, therefore, one of your most important tasks in helping your clients will be working hard to communicate your empathy, interest, warmth, and caring, and to make certain that your clients are aware of your feelings for them.

Gestalt Therapy *Gestalt therapy* is similar to Rogers' client-centered approach because it emphasizes client self-awareness and growth facilitated by participation in an authentic counseling relationship. Gestalt therapy is experiential and stresses client exploration and integration of thoughts, emotions, and behaviors through awareness of the here-and-now (Corey, 2001a, 2001c). The client and the counselor work to compare their individual phenomenological perspectives. They focus on exploring their differing viewpoints, with the goal of increasing client awareness, insight, acceptance, and responsibility (Corey, 2001a; Yontef & Jacobs, 2000).

Gestalt therapy places more emphasis on specific counseling techniques, called *experiments* or *exercises*, than do the other humanistic/existential approaches. One of the Gestalt techniques that we have found to be very useful to counselor interns involves helping your client to identify and focus on the feeling (*e.g.*, anger) that is beneath his or her nonverbal behavior (*e.g.,* clenched fists), and then asking your client to "Stay with the feeling." As the client experiences the feeling fully, he or she often gains self-awareness and becomes more able to explore important issues.

Another Gestalt technique you may find useful during your internship is known as the *empty chair*. Here, clients imagine someone, such as a parent or spouse, to be sitting in a nearby chair. Clients then say all the things to the empty chair that they have never been able to say to the real person. This directed verbalization often is a cathartic experience for the client and leads to new understandings for both client and counselor.

Many Gestalt techniques will be useful to you during your internship, because they will help you attend closely to the significance of your clients' nonverbal behaviors as well as their use of language in relation to their thoughts and feelings. As a counselor intern, you will be able to understand your clients more fully and grasp the depths of their feelings and experience by attending carefully to their nonverbal behaviors and cues. For example, noticing your clients' muscle tension, body posture, and hand and facial movements provides you with a great deal of information. In addition, you will be able to understand and empathize with your clients' perceptions of themselves and their worlds by listening closely to the words they select to express themselves. For example, the client who says, "He makes me so depressed," feels less powerful and assumes less responsibility than the client who says, "I feel depressed when he does that."

Corey (2001a, 2001b) cautions that the humanistic/existential therapies focus on the present moment to such a great extent that the influence of the client's developmental history and the power of the past too often may be discounted. In addition, he suggests that the emphasis on emotions results in less attention to the importance of cognitive or intellectual factors in understanding clients.

Postmodern Therapies

Theoretical orientations known as *postmodern therapies* share an emphasis on the social construction of meaning and the importance of language in determining the way people think, feel, and behave.

Solution-Focused Brief Therapy   *Solution-focused brief therapy*, also known as *SFBT*, was introduced by Steve de Shazer and Insoo Kim Berg in the early

1980s and has grown tremendously in popularity since that time (de Jong & Berg, 2002). SFBT is unique in that it focuses exclusively on constructing solutions to client problems, and asserts that identifying and exploring problems is unimportant and often hinders progress (de Jong & Berg, 2002). Therapists ask clients to envision a more satisfying or enjoyable future, and then work with them to develop an awareness of client strengths and resources that can be used to build solutions to transform this vision into reality (de Jong & Berg, 2002). The therapeutic relationship becomes the context of change, and client and therapist co-construct reality through the use of language, primarily "change talk" (Gingerich, 2002). Reviews of the outcome research indicate that SFBT has a broad range of applicability and is useful in a wide variety of settings and with diverse populations (Gingerich & Eisengart, 1999, 2000, 2001). There is no specified number of sessions for SFBT, although clients rarely require more than six visits.

Several techniques are used to foster solution-building in SFBT. The first of these is the *Miracle Question*, in which the therapist asks clients to imagine that a miracle will occur when they go to sleep tonight, and that this miracle will solve all their problems. The therapist then asks, "What will you be doing the morning after this miracle happens? What will be different?" The therapist continues with questions designed to elicit the client's behavioral goals for therapy.

A second technique is called *looking for exceptions*. Here, the therapist asks the client if things have been even a little bit better at any time or if part of the miracle has ever occurred. SFBT assumes that the solution, or at least a portion of it, is already present at least part of the time (de Jong & Berg, 2002). When the client and therapist identify an instance when part of the solution has been present, they work together to determine what factors have contributed to the solution and how these can be amplified or increased in frequency (Gingerich, 2002). For example, the woman who says her children "fight constantly" may be surprised when she thinks back and recalls that last week they played together for one hour without fighting at all.

A third SFBT technique is *using compliments* (de Jong & Berg, 2002). This refers to giving the client positive feedback along with any intervention. The compliment should highlight a client strength that the therapist believes can be useful in promoting constructive client change. This technique serves to empower clients by helping them recognize their own strengths. To be effective, compliments must be *accurate*, *believable*, and *constructive* (Gingerich, 2002).

A fourth technique is the use of *scaling questions*. Clients are asked to rate their progress on a scale of 1 to 10, with 1 being the situation as it was when they decided to make their first appointment for counseling, and 10 being the situation the day after the miracle occurs. SFBT researchers have found that clients often experience *pre-therapy improvement* and begin to feel better as soon as they have called to schedule their first appointment for therapy (de Jong & Berg, 2002). Clients, and therapists, too, are frequently surprised when clients who say they feel very depressed at the first session rate their mood as a 6 or 7 at that session. This use of scaling questions instills hope and enables the client to view his or her situation more realistically.

Another way to make use of scaling questions is to ask clients to indicate where they are at the present time in terms of reaching a specific goal, and what would be different if they were one number higher. For example, suppose a client's goal is to begin a regular exercise routine. If the client reports being at a number 4 with regard to the likelihood of beginning to exercise regularly, the therapist would ask, "What would you be doing differently if you were at a number 5?" Often, this helps the client and therapist understand how to break the solution down into small, manageable steps. Finally, scaling questions enable client and counselor to measure progress in a concrete way, enhancing the client's feelings of mastery and competency.

Narrative Therapy *Narrative therapy* is based on the premise that people use stories to guide their lives, organize information about themselves, and make sense of their experiences (Payne, 2000; Sween, 1999). Narrative therapy evolved from family therapy traditions, and stories are viewed in the context of family history, culture, and society (Freedman & Combs, 1996). An underlying principle of narrative therapy is that the client is respected as a valuable human being and is not considered to be the problem; client and counselor look at the problem as a separate entity that needs to be solved (Sween, 1999).

Narrative therapists search for the stories that illuminate the forces that have shaped clients' lives and possibly have caused setbacks. They also search for the forgotten stories that provide evidence of the client's strengths and abilities (Sween, 1999). People's identities are constructed in part by the way they are viewed by others (Freedman & Combs, 1996). When we understand the myriad stories that direct our lives, we can identify those that have sustained our problems and we can write alternate stories that reflect new identities (Payne, 2000). Together, counselor and client can work to construct a new, more positive reality by rewriting and retelling the client's life stories (Eron & Lund, 1998).

Dialectical Behavior Therapy *Dialectical behavior therapy*, commonly known as *DBT*, was developed by Marsha Linehan as a treatment for clients with chronic suicidal thoughts and persistent self-mutilating behaviors. It is widely used to treat clients with borderline personality disorder, and it is the only treatment for this complex disorder that is supported with solid outcome research evidence (Koerner & Linehan, 2000; Linehan, Cochran, & Kehrer, 2001). More recently, studies have indicated that DBT is also a valid treatment for binge-eating disorder (Telch, Agras, Stewart, & Linehan, 2001), bulimia nervosa (Safer, Telch, & Agras, 2001), and impulsive self-injurious behavior (Ivanoff, Linehan, & Brown, 2001).

DBT is anchored in postmodern and cognitive-behavioral approaches, as well as in Zen principles. Linehan (Linehan, Cochran, & Kehrer, 2001) writes that therapists must be empathic, but must balance this empathy with a quality she terms "irreverence." She acknowledges that clients with borderline personality disorder are in extreme pain, but she states that they still need to learn skills to manage their own problems. Because borderline personality disorder is so difficult to treat, Linehan asserts that therapists need support to prevent burn out, and she utilizes a therapist support group in her own work (Linehan, Cochran, & Kehrer, 2001).

According to Linehan (Linehan, Cochran, & Kehrer, 2001), clients with borderline personality disorder are suffering from dysregulation in five areas, including emotions, interpersonal relationships, behaviors, self, and thinking. She writes that these deficits may be caused by a physiological hyper-reactivity, a biological problem with affect management, and exposure to an invalidating or traumatic environment. To address their problems, Linehan's DBT protocol includes, in the following order:

- Contracting for no suicidal or self-harming actions
- Decreasing behaviors that interfere with therapy
- Decreasing behaviors that interfere with the client's quality of life
- Increasing self-care, distress tolerance, and interpersonal and coping skills
- Decreasing symptoms of posttraumatic stress disorder (PTSD)
- Increasing self-esteem
- Encouraging the attainment of individual goals (Linehan, Cochran, & Kehrer, 2001)

ADJUNCT TREATMENT MODALITIES

Treatment modality refers to the configuration or structure of counseling used to help the client or clients. During your internship, the treatment modality you will probably use most often will be *individual therapy*, in which you will be talking with one client at a time. Depending on the population you serve, you may also have an opportunity to use adjunct treatment modalities, including *group therapy, bibliotherapy,* and *art therapy*.

Group Therapy

You will most likely be asked to facilitate or co-facilitate at least one group as part of your internship training. Group counseling is used in a wide range of settings, with diverse populations, and for varied purposes. Groups generate powerful, potentially therapeutic interpersonal forces that may be used for education, support, and prevention or remediation of psychological issues (Corey & Corey, 1997). The general goals of group counseling consist of learning to trust self and others, recognizing commonalities of problems and needs, achieving greater self-awareness, increasing self-confidence and self-acceptance, finding alternative ways of managing problems and resolving conflicts, becoming more aware of one's responsibility toward self and others, ventilation of feelings, instillation of hope, and improved interpersonal skills (Corey, 2000; Corey & Corey, 1997; Yalom, 1995).

When you facilitate groups during your internship, you can focus your interventions toward accomplishing these goals, and helping individual members to use the group's healing forces. In concrete terms, that means:

- Ensuring a safe place in the group by protecting members physically and psychologically
- Providing appropriate structure and clear rules so that a therapeutic framework is maintained
- Encouraging direct communication and interaction among group members
- Modeling active listening skills
- Providing positive feedback when group members take risks, work hard, show increasing self-awareness, offer support for others, and so forth
- Being aware of the developmental stages of group process
- Asking questions such as "How many others have felt that way?" or "How many others have had a similar experience?"
- Helping members become aware of and express their current feelings concerning group interaction

Bibliotherapy

Bibliotherapy involves using books or other reading material to help your clients. A book may be of the popular "self–help" variety, a booklet prepared especially for clients with a particular problem, a professional resource that presents relevant information without using too much jargon, or information downloaded from a reliable Web site on the Internet. You may suggest that your client read during the time between your sessions for "homework," or you may read an appropriate passage together during the session. Some groups are structured so that members take turns reading aloud, followed by a discussion of personal reactions to each section.

In each situation, we suggest that you first read the material yourself to be sure that it is accurate and appropriate for your client's unique concerns and current stage in the counseling process. In addition, you can use bibliotherapy most effectively by helping your clients explore their own thoughts and feelings about the reading material. We have noticed that quite often bibliotherapy not only provides clients with factual information, but also offers some of the same therapeutic factors we mentioned in the section on group therapy, including universality, guidance, identification, self-understanding, and instillation of hope.

Art Therapy

Neither you nor your client needs to have any special artistic talents to use art during your counseling sessions. *Art therapy* is an alternative, nonverbal means of communication, which you can use in a helpful way with clients during your internship. We tell clients that, "Art helps us say things that sometimes just can't be put into words." When using art with clients, keep in mind that any artwork produced during your counseling session is likely to be a very important, intensely personal representation of the client's sense of self or inner world, because feelings and issues are being stirred up by the treatment situation. Make sure to demonstrate your interest, caring, and warm acceptance of the artwork. Also, take time to explore the client's feelings and thoughts by talking about the artwork.

We suggest using simple colored markers and a 9-by-12-inch pad of white paper during your internship. These are easy to carry with you, do not require any special space or preparation, and are nonthreatening and easy for most clients to use. You may find it helpful to explain as you offer the art materials that, "This is about feelings, not about the art being good or bad." Many clients are hesitant to begin and may need some encouragement or direction. If this is the case, you can offer one or two specific suggestions about what to draw. Possible topics include: an event from the past, a dream, the client's family of origin, how the client is feeling right now, a wish, a favorite activity, a fear, an important person in the client's life, a self-portrait, a dream, or a monster.

When a client draws an upsetting or frightening picture, you can help him or her experience a sense of power and control by actually changing the drawing to make it safer or less scary. For example, one young client became anxious and afraid after drawing a picture of a large dog that had attacked and bitten her. She felt safer when she drew a strong cage around the dog, and added a lock on the cage door. Another client who had been mistreated by someone experienced a sense of relief when she chose to draw a picture of him, rip it up, and throw the pieces of paper in the trash.

Art is a powerful therapeutic tool, and we have found it to be helpful because it:

- Provides a cathartic experience; strong emotions are released as art is produced.

- Increases the connection between counselor and client, as the artwork is shared and discussed.

- Helps increase client self-esteem. The counselor values and accepts the client's artwork and, in so doing, demonstrates that the client is also worthy of being valued and accepted.

- Is empowering for the client, who is engaged in direct manipulation of his or her environment while having control of the creative process.

CONCEPTUALIZING CLIENT CONCERNS

A Biopsychosocial Perspective:
The Individual and the Environment

The first step in choosing an appropriate treatment modality or therapeutic intervention involves working to understand your client from a biopsychosocial perspective. We encourage counselor interns to take note of genetic, biological, and intrapsychic forces in relation to external influences of the family, society, and world (Dacey & Travers, 1999). There is always a dynamic interplay between the client and the situation. As a counselor intern, it is important to use a comprehensive perspective, considering the client as an individual interacting with a complex, multidimensional environment.

Consider your client's problems, behaviors, and concerns in context at all times, and keep in mind that seemingly abnormal behaviors may be very normal responses to an abnormal or highly stressful situation. Your client's problematic symptoms can often be reframed as his or her "solution" to coping with internal and/or external stressors. Although the "solution" may have made sense at one time, it no longer works in the client's best interests. Together, you and your client can begin to understand the etiology and the purpose of the symptom in order to replace it with a more acceptable coping skill.

A Developmental Perspective:
The Significance of Life Stages

During your internship, you also will find it useful to assess your client's personality traits, concerns, and behaviors within a framework provided by personality theorists and developmental psychologists. Clients often require help in negotiating critical turning points in the life span or in dealing with difficult developmental tasks (Corey, 2001c). Knowledge of developmental stages allows you to understand what types of problems are likely to arise at each specific phase of your client's life cycle, so that you can be most helpful. Awareness of developmental norms enables you to assist your client back onto a healthy developmental pathway.

In addition, a thorough knowledge of developmental stages provides a template for determining the relative health or pathology of your client (Gabbard, 2000). Behavior that is appropriate and acceptable at one stage of life may be troublesome at another stage. For example, a 14-year-old boy who tells us he has never dated because he is too anxious to ask a girl to go out presents a vastly different clinical picture than a 35-year-old man who has the same problem. As counselors, we are not concerned when a 2-year-old expresses disagreement by having a temper tantrum, but we see a red flag if a 20-year-old engages in such behavior. Similarly, we are not worried about the 80-year-old individual who spends hours sitting and reminiscing about the past, although we may be concerned about the 40-year-old person who chooses to do so.

A Multicultural Perspective: Appreciating Diversity

Culture implies a set of prescribed values, norms, traits, communication styles, and behavioral patterns that constitute a unique cultural reality and provide a sense of identity for members of the group (Lee, 1997). Cultural factors may include race, ethnicity, social class, gender, age, religious affiliation, and sexual orientation (Gonzales, Griffith, & Ruiz, 2001). Cultural identity colors the way the individual perceives the self and the world and provides meaning and structure to everyday activities, interpersonal relationships, and social behaviors (Gibbs & Huang, 1998). Culture also determines the way people experience, express, and cope with distress, as well as how, where, and from whom they seek help (Tseng & Streltzer, 2001). Thus, cultural factors permeate every aspect of the counseling process.

Sensitivity and attention to diverse cultures will enable you to establish greater rapport with your client, enhance the therapeutic relationship, understand the client's experiences and values more fully, and communicate more effectively. For counselors, developing sensitivity to diversity entails becoming aware of one's own culturally based values, beliefs, and norms; acknowledging that these may be very different from the client's; and accepting the client's values, beliefs, and norms in a nonjudgmental way (Tseng, 2001). Counselors must expand their cultural knowledge base and adjust their therapeutic skills, so that they will be more able to understand their clients' concerns in diverse cultural contexts (Tseng & Streltzer, 2001). Being aware of culturally based concepts of sickness and health, as well as culturally embedded behavioral norms, allows you to work with clients to set goals that are both personally and culturally appropriate.

Counselors need to be cognizant of the fact that certain values we may take for granted in Western, androcentric cultures may be antithetical to the values of other cultures. For example, Western cultures place emphasis on an individual orientation rather than on a group orientation, whereas Eastern cultures may view the group as more important than the individual (Sue & Sue, 1999). Other values that counselors need to examine in light of potential cultural differences include degree of emotional, verbal, or behavioral expressiveness; worth of insight and self-examination; comfort with openness and self-disclosure; importance of scientific thought; and acceptance of ambiguity (Sue & Sue, 1999). Attention to these cultural factors, as well as to patterns of verbal and nonverbal communication, will help you modify traditional treatment approaches and interventions to be more congruent with your clients' diversity needs (Gonzales, Griffith, & Ruiz, 2001; Sue & Sue, 1999; Tseng & Streltzer, 2001).

CONCLUDING REMARKS

The counselor's relationship with the client determines the helpfulness of counseling, and the client's awareness of the counselor's empathy is the most important aspect of the therapeutic relationship (Kahn, 1997). Therefore, during your internship, we hope that you will value yourself as your most powerful means of helping your client. As a beginning counselor, your theoretical orientation, the specific treatment modalities you use, and your therapeutic interventions are not as important as your caring approach, your sincere efforts to understand your client, and your ability to convey to your client your empathy, respect, and desire to be helpful.

In this chapter, we have included an overview of some of the concepts and philosophies underlying the major theoretical orientations, techniques, and interventions arising from these theoretical approaches, and treatment modalities we feel are most relevant to you as a counselor intern. We have limited our choices not only with regard to the constraints of the time and space of one chapter of a single book, but also with respect to your level of training as an intern. As you gain clinical experience, become more comfortable interacting with clients, encounter more difficult problems, refine your personal counseling style, and function with more autonomy in your counseling career, we encourage you to enrich your repertoire of treatment modalities and counseling methods to help your clients reach their highest levels of functioning.

REFERENCES

BALDWIN, M. (Ed.). (2000). *The use of self in therapy.* New York: Haworth Press.

BECK, A., & WEISHAAR, M. (2000). Cognitive therapy. In R. Corsini & D. Wedding (Eds.), *Current psychotherapies* (6th ed., pp. 241–271). Itasca, IL: Peacock.

COREY, G. (2000). *Theory and practice of group counseling* (5th ed.). Pacific Grove, CA: Brooks/Cole.

COREY, G. (2001a). *Manual for theory and practice of counseling and psychotherapy.* Pacific Grove, CA: Brooks/Cole.

COREY, G. (2001b). *The art of integrative counseling.* Australia: Brooks/Cole.

COREY, G. (2001c). *Theory and practice of counseling and psychotherapy* (6th ed.). Pacific Grove, CA: Brooks/Cole.

COREY, G., COREY, M. S., & CALLAHAN, P. (1998). *Issues and ethics in the helping professions* (5th ed.). Pacific Grove, CA: Brooks/Cole.

COREY, M., & COREY, G. (1997). *Groups: Process and practice* (5th ed.). Pacific Grove, CA: Brooks/Cole.

DACEY, J., & TRAVERS, J. (1999). *Human development across the lifespan* (4th ed.). Dubuque, IA: Brown.

DE JONG, P. & BERG, I. K. (2002). *Interviewing for solutions* (2nd ed.). Pacific Grove, CA: Wadsworth/Thompson.

DOYLE, R. (1998). *Essential skills and strategies in the helping process* (2nd ed.). Pacific Grove, CA: Brooks/Cole.

ELLIS, A. (2000). Rational emotive behavior therapy. In R. Corsini & D. Wedding (Eds.), *Current psychotherapies* (6th ed., pp. 168–204). Itasca, IL: Peacock.

ERON, J. B., & LUND, T. W. (1998). *Narrative solutions on brief therapy.* New York: Guilford.

FREEDMAN, J., & COMBS, G. (1996). *Narrative therapy: The social construction of preferred realities.* New York: W. W. Norton.

GABBARD, G. (2000). *Psychodynamic psychiatry in clinical practice* (3rd ed.). Washington, DC: American Psychiatric Press.

GIBBS, J., & HUANG, L. (1998). A conceptual framework for the psychological assessment and treatment of minority youth. In J. Gibbs & L. Huang (Eds.), *Children of color* (2nd ed., pp. 1–32). San Francisco: Jossey-Bass.

GINGERICH, W. J. (2002). A day-long solution-focused brief therapy workshop. Retrieved on May 5, 2002, from the Wallace J. Gingerich Web site: www.gingerich.net/SFBT/workshop.htm

GINGERICH, W. J., & EISENGART, S. (1999). *What is the evidence? A review of the outcome research on solution-focused brief therapy.* Paper presented at the European Brief Therapy Association, Carlisle, UK.

GINGERICH, W. J., & EISENGART, S. (2000). Solution-focused brief therapy: A review of the outcome research. *Family Process, 39*(4), 477–498.

GINGERICH, W. J., & EISENGART, S. (2001). *Solution-focused brief therapy: What is the empirical support?* Paper presented at the American Association for Marriage and Family Therapy, Nashville, Tennessee.

GONZALES, C. A., GRIFFITH, E. E. H., & RUIZ, P. (2001). Cross-cultural issues in psychiatric treatment. In G. O. Gabbard (Ed.), *Treatment of psychiatric disorders* (3rd ed., Vol. 1, pp. 47–67). Washington, DC: American Psychiatric Publishing, Inc.

HOLMES, J. (1999). Ethical aspects of the psychotherapies. In S. Bloch, P. Chodoff, & S. A. Green (Eds.), *Psychiatric ethics.* London: Oxford University Press.

IVANOFF, A., LINEHAN, M. M., & BROWN, M. (2001). Therapy for self-injurious behaviors. In D. Simeon (Ed.), *Self-injurious behaviors: Assessment and treatment* (pp. 149–173). Washington, DC: American Psychiatric Press.

KAHN, M. (1997). *Between therapist and client: The new relationship* (2nd ed.). New York: Freeman.

KOERNER, K., & LINEHAN, M. M. (2000). Research on dialectical behavior therapy for patients with borderline personality disorder. *Psychiatric Clinics of North America, 23*(1), 151–167.

KOHUT, H. (1977). *The restoration of the self.* New York: International Universities Press.

LEE, C. (Ed.). (1997). *Multicultural issues in counseling: New approaches to diversity.* Alexandria, VA: American Counseling Association.

LINEHAN, M. M., COCHRAN, B. N., & KEHRER, C. A. (2001). Dialectical behavior therapy for borderline personality disorder. In E. Barlow (Ed.), *Clinical handbook of psychological disor-*ders: *A step-by-step treatment manual* (3rd ed., pp. 470–522). New York: Guilford Press.

MESSER, S. B. (2001). What makes brief psychodynamic therapy time efficient. *Clinical Psychology: Science and Practice, 8*(1), 5–22.

MONK, G., WINSLADE, J., & CROCKET, K. (Eds.). (1997). *Narrative therapy in practice: The archaeology of hope.* San Francisco: Jossey-Bass.

PAYNE, M. (2000). *Narrative therapy: An introduction for counsellors.* London: Sage Publications.

RASKIN, N., & ROGERS, C. (2000). Person-centered therapy. In R. Corsini & D. Wedding (Eds.), *Current psychotherapies* (6th ed., pp. 133–167). Itasca, IL: Peacock.

SAFER, D. L., TELCH, C. F., & AGRAS, W. S. (2001). Dialectical behavior therapy for bulimia nervosa. *American Journal of Psychiatry, 158*(4), 632–634.

SATIR, V. (1987). The therapist story. In M. Baldwin & V. Satir (Eds.), *The use of self in therapy* (pp. 17–26). New York: Haworth.

SHEA, S. (1998). *Psychiatric interviewing: The art of understanding* (2nd ed.). Philadelphia: Saunders.

SIEGEL, A. M. (1996). *Heinz Kohut and the psychology of the self.* New York: Routledge.

SUE, D. W., & SUE, D. (1999). *Counseling the culturally different* (3rd ed.). New York: John Wiley & Sons, Inc.

SWEEN, E. (1999). *The one minute question: What is narrative therapy?* Retrieved April 30, 2002, from the Jung Page Web site: www.cgjungpage.org/articles/sween1.html

TELCH, C. F., AGRAS, W. S., STEWART, W., & LINEHAN, M. M. (2001). Dialectical behavior therapy for binge-eating disorder. *Journal of Consulting and Clinical Psychology, 69*(6), 1061–1065.

TSENG, W. S. (2001). Overview. In W. S. Tseng and J. Streltzer (Eds.), *Culture and psychotherapy: A guide to clinical practice* (pp. 3–12). Washington, DC: American Psychiatric Press.

TSENG, W. S., & STRELTZER, J. (2001). Integration and conclusions. In W. S. Tseng and J. Streltzer (Eds.), *Culture and psychotherapy: A guide to clinical practice* (pp. 265–278). Washington, DC: American Psychiatric Press.

YALOM, I. (1995). *Theory and practice of group psychotherapy* (4th ed.). New York: Basic Books.

YONTEF, G., & JACOBS, L. (2000). Gestalt therapy. In R. Corsini & D. Wedding (Eds.), *Current psychotherapies* (6th ed., pp. 303–339). Itasca, IL: Peacock.

BIBLIOGRAPHY

CHIAFERI, R., & GRIFFIN, M. (1997). *Developing fieldwork skills*. Pacific Grove, CA: Brooks/Cole.

KOTTLER, J., & BROWN, R. (1996). *Introduction to therapeutic counseling* (4th ed.). Pacific Grove, CA: Brooks/Cole.

SELIGMAN, L. (1998). *Selecting effective treatments: A comprehensive, systematic guide to treating mental disorders* (2nd ed.). Alexandria, VA: American Counseling Association.

URSANO, R., SONNENBERG, S., & LAZAR, S. (1998). *Concise guide to psychodynamic psychotherapy: Principles and techniques in the era of managed care.* Washington, DC: American Psychiatric Press.

8

Understanding
Psychotropic Medications

The word *psychotropic* is derived from two Greek words: *psyche,* meaning *mind,* and *tropos,* meaning *to turn* (*Webster's Unabridged Dictionary,* 2001). Thus, psychotropic medications are drugs that *turn (or influence) the mind,* with the goal of alleviating distressing mental and emotional symptoms. *Psychopharmacology* is the study of interrelationships among psychotropic medications, mental disease, and brain chemistry, structure, and function (Stahl, 1996). *Psychopharmacotherapy* (often shortened to *pharmacotherapy*) refers to the use of psychotropic medications to treat mental and emotional disorders.

In the early 1950s, a revolution in mental health care began with the discovery that thorazine, a medication originally developed as an antihistamine, also reduced symptoms of psychosis in patients with schizophrenia (Stahl, 1996). Since that time, rapidly expanding neuroscience research has produced a profusion of new, safer, and highly effective psychotropic medications (Gabbard, 2001). These medications have decreased the frequency and length of psychiatric hospitalizations, alleviated the severity of symptoms in serious mental illnesses such as schizophrenia and bipolar disorder, and provided relief and hope to millions of persons suffering from depression and anxiety disorders (Delgado & Gelenberg, 2001; Gabbard, 2001; Kane & Malhotra, 2001; Lydiard, Otto, & Milrod, 2001).

WHY SHOULD COUNSELORS STUDY PSYCHOTROPIC MEDICATIONS?

The development of safer and more efficient psychotropic medications, in conjunction with extensive dissemination of information via the Internet, popular books, and print and broadcast media, has resulted in greater acceptance and widespread use of these medications. Furthermore, because psychotropic medications provide cost-effective and rapid reduction of symptoms, they have become the predominant treatment modality for psychiatric disorders (Gabbard, 2001). They are prescribed not only by psychiatrists, but also by pediatricians, family practice doctors, and other medical specialists, and now comprise a large proportion of the most commonly prescribed drugs in the world (National Institute for Health Care Management, 2002). In fact, antidepressants were the top-selling category of all prescription medications in 2001, with retail sales totaling more than 10 billion dollars in the United States (National Institute for Health Care Management, 2002). Because psychotropic medications are so prevalent, counselors and counselor interns need to develop familiarity with them, as well as skills in interfacing with individuals who are taking, or who may need referral for, psychotropic medications.

MECHANISM OF ACTION: HOW DO PSYCHOTROPIC MEDICATIONS WORK?

Before we begin our discussion of the mechanism of action of psychotropic medications, we will set the stage by providing a few basic facts about brain function. All human thoughts, feelings, and behaviors originate as chemical and electrical transmission of information in our brains. *Neurons* are brain cells that are specially adapted to communicate with one another by passing along this information. There are approximately 100 billion neurons in the human brain, arranged in interconnected *neural circuits*, or "brain wiring" (Iacono, 2000). *Neurotransmitters* are chemicals produced in the brain that transmit information across the spaces between individual neurons. These spaces, known as *synapses*, facilitate complex chemical and electrical connections in the neural circuits (*Neurotransmitters*, 2002). Three of the most important neurotransmitters are *serotonin, noradrenaline*, and *dopamine* (Korn, 2001; *Neurotransmitters*, 2002). Only about 20% (that is, about 28 billion) of all neurons have the capacity to produce neurotransmitters, but these special neurons branch very widely and are interconnected to all the other neurons, thus supervising transmission of information in the brain (Korn, 2001).

Transmission of information begins when molecules of a neurotransmitter are released by a neuron that is sending information, known as a *presynaptic neuron*. These molecules travel across the synapse and bind to specific *receptor sites* in a neuron that is receiving information, known as a *postsynaptic neuron* (*Neurotransmitters*, 2002). When neurotransmitter molecules successfully bind to various receptor sites, an electrical charge in the postsynaptic neuron is activated, completing communication between the neurons and giving rise to a wide variety of complex brain functions, including memory, learning, creation of new ideas, emotions, sight, hearing, taste, tactile and temperature sensations, and behavior (*Neurotransmitters*, 2002; Stahl, 1996). Not all the neurotransmitter molecules that are released into the synapse have an opportunity to bind to receptor sites. In a process known as *reuptake*, some of the molecules are reabsorbed by the presynaptic neuron, so that they do not stay in the synapse long enough to bind to receptor sites.

Because most psychiatric disorders are associated with an abnormality in the levels or action of neurotransmitters, the majority of psychotropic medications work by influencing either the reuptake or receptor-binding processes of one or more neurotransmitters (Kramer, 2002a; Stahl, 1996). For example, some psychotropic medications used to treat depression block the reuptake of serotonin or noradrenaline, or both, increasing the time that these two neurotransmitters remain in the synapses and altering the way they bind to receptor sites (Korn, 2001; Stahl, 1996). Similarly, some of the newer medications used to reduce the psychotic symptoms of schizophrenia regulate reuptake of dopamine and noradrenaline (Kramer, 2002a).

PSYCHOTROPIC MEDICATION FOR CHILDREN AND ADOLESCENTS

The use of psychotropic medications to treat emotional and behavioral problems in children has risen dramatically in the past 10 years, with a substantial increase in the numbers of prescriptions written for children between two and four years old (Coyle, 2000; Rushton & Whitmire, 2001; Zito, Safer, dosReis, Gardner, Boles, & Lynch, 2000; Zito, Safer, dosReis, Magder, Gardner, & Zarin, 1999). Some research supports the safety of short-term use of psychotropic medications to treat children and adolescents (National Institute of Mental Health, 2001), but there are very few follow-up studies of the long-term effects of psychotropic medications on youngsters (Coyle, 2000; Martin, Kaufman, & Charney, 2000). In fact, one medication commonly prescribed for preschoolers (*i.e.*, Ritalin) carries a warning label that it is not safe for children under six years of age (Coyle, 2000; Martin, Kaufman, & Charney, 2000).

The question of whether to give psychotropic medications to children is very complex and presents a treatment dilemma. More than 10% of children and adolescents suffer from mental and emotional disorders severe enough to impair functioning (National Institute of Mental Health, 2000). These disorders may cause serious problems in all realms of life, including relationships with family and peers, ability to concentrate and learn at school, development of social skills, and self-esteem. Although psychotropic medications often provide effective treatment and can relieve a child's or adolescent's suffering, many experts have valid concerns about the potential deleterious effects of psychotropic medications on the developing brain (Coyle, 2000). To make informed decisions about these medications for children and adolescents, we need to be aware of certain facts.

First, we must be cognizant of the fact that human brain development begins *in utero* and continues throughout childhood and adolescence, into the early 20s (Casey, Giedd, & Thomas, 2000; Epstein, 2001; National Institute of Mental Health, 2001). Second, we must revisit the role of neurotransmitters. In a previous section of this chapter, we learned that neurotransmitters facilitate transmission of information between neurons in the adult brain. However, neurotransmitters serve an additional, critical function in the development of the human brain, both *in utero* and in childhood, by controlling the growth and shape of neurons and guiding the arrangement of neural circuits (Herlenius & Lagercrantz, 2001). Substantial research on animals has shown that disturbance of neurotransmitters during brain development tends to alter the growth of neurons and to disrupt the brain's wiring, causing long-term cognitive, emotional, and behavioral deficits (El-Mallakh, Peters, & Waltrip, 2000; Gabbard, 2001). As we noted earlier, psychiatric disorders are associated with a disturbance in the levels or action of neurotransmitters, and psychotropic medications work by influencing neurotransmitters. Should we administer psychotropic medications to children and adolescents to try to correct the imbalance of neurotransmitters associated with a psychiatric disorder? Will the medication's influence on the developing brain cause more damage than benefit? More research is needed to elucidate the possible long-term effects of both psychotropic medications and psychiatric disorders on the brains of children and adolescents. Until more scientific evidence is available, the potential risks and benefits of medication versus the potentially harmful biological and psychosocial effects of the psychiatric disorder must be carefully evaluated for each child or adolescent, and the family must be included in the decision-making process.

HOW ARE PSYCHOTROPIC MEDICATIONS GIVEN?

Psychotropic medications are manufactured and prescribed in several different forms. Most commonly, they are prescribed as a capsule or pill, taken two or three times each day, or once before bed, in the case of sleep medications. Some medications are available in extended-release form, so that one capsule or pill releases a continuous dose of medication over a 12- or 24-hour period of time. Certain medications (*e.g.*, Prozac) are available in a week-long extended-release form, so that the client needs to take a pill only once each week. Medications are also available as liquids, which is especially helpful in situations where dosages need to be titrated carefully or when medications must be administered to young children or to others who are unable to swallow pills. Other medications (*e.g.*, Depakote) are available in "sprinkles" form, so that they may be mixed with applesauce or other foods to make them more palatable. Injections may be used when rapid onset of action is needed, especially when antianxiety or antipsychotic medications are indicated to calm agitated persons. Some antipsychotics are available in the form of a long-acting injection, which is helpful for patients with serious mental illnesses who have not been compliant with taking their pills.

COMBINED TREATMENT: PSYCHOTHERAPY AND PSYCHOTROPIC MEDICATIONS

Combined treatment refers to a situation in which the client sees a counselor for psychotherapy and a physician for medication management. Due to a growing awareness of the complex, multicausal etiology of psychiatric disorders, as well as cost-containment efforts of third-party payers, the integration of psychotherapy and pharmacotherapy has become a widely accepted practice (Riba & Balon, 2001). Abundant research evidence indicates that combined treatment is almost always more useful than either psychotherapy or psychotropic medications alone (Gabbard, 2001; Kay, 2001; Schatzberg & Nemeroff, 2001). An integrated approach has been shown to be effective for treating alcoholism, anxiety disorders, attention deficit-hyperactivity, borderline personality disorder, chronic pain, eating disorders, mood disorders, premenstrual dysphoric disorder, schizophrenia, sexual dysfunction, sleep disorders, tic disorder, and many other problems (Kaplan & Sadock, 2001; Kay, 2001; Pollack, Otto, & Rosenbaum, 1996; Reid, Balis, & Sutton, 1997).

Benefits of Combined Treatment

Psychotropic medications and psychotherapy work synergistically, providing the most comprehensive and efficient treatment for the majority of mental and emotional disorders (Feldman & Feldman, 1997a; Kay, 2001; Schatzberg & Nemeroff, 2001). Each treatment approach has strengths and disadvantages, but using pharmacotherapy and psychotherapy concurrently enhances the benefits and overcomes the limitations of each one (Feldman & Feldman, 1997b). Combined treatment increases the probability of a good response to treatment, because it is more likely that the client's problems will be adequately addressed with a dual approach (Hollon & Fawcett, 2001). Psychotropic medications provide relief from problematic symptoms, such as anxiety, obsessive thoughts, fatigue, low energy, difficulty concentrating, or hallucinations, which may impede effective participation in the therapy process (Feldman & Feldman, 1997b; Gabbard, 2001). Psychotherapy often improves clients' compliance with their medication regimens and increases their ability to tolerate side effects (Hollon & Fawcett, 2001; Riba & Balon, 2001). Thus, combined treatment frequently shortens the overall length of treatment and reduces relapse rates (Gabbard, 2001; Kay, 2001; Riba & Balon, 2001).

There are other benefits of combined treatment, as well. Disruptions in treatment may be reduced when there are two clinicians involved with the client (Kay, 2001). This may be especially helpful in working with clients who feel abandoned or become upset when the clinician goes on vacation or is otherwise unavailable. In addition, if a conflict, misunderstanding, or transference issue derails the relationship between the client and one of the clinicians, the other clinician can be available to maintain the client's engagement in therapy and help heal the problem (Feldman & Feldman, 1997b). Moreover, clinicians can benefit from the support of and consultation with another colleague who is familiar with the case.

Responsibilities of Counselors in Combined Treatment

Because the structure of the mental healthcare system reduces the frequency of client contacts with more costly providers (such as psychiatrists), a counselor may be the professional helper with whom the client interacts most often. Counselors have an ethical responsibility to become knowledgeable about all treatment options and to share this knowledge with their clients; thus, counselors need to become familiar with psychotropic medications (Ingersoll, 2001).

We agree with Ingersoll's (2001) description of the nonmedical professional's role concerning psychotropic medications as collaborator, information broker, and support person. Indeed, when counseling clients who are taking (or who may need to be referred for) psychotropic medications, you will collaborate with the psychiatrist or other prescribing physician and also with the client; you will provide information to the client about the need for referral and combined treatment; and you will provide ongoing support to the client regarding medications. These professional tasks require highly specialized knowledge and clinical skills, so you should always consult carefully with your site supervisor if you are concerned about a client's potential need for medication referral or a problem with a client who is already taking psychotropic medications.

We cannot stress strongly enough that counselors should *never* give advice concerning what medications the client may need, how much medication to take, or when to take it. As counselors, we are not trained or licensed to practice medicine. We should encourage compliance by reminding clients to take their medications *exactly as their doctor has prescribed*, by directing clients to call their physicians when they have questions, and by calling the physician ourselves (with the appropriate signed release of information) if we have concerns.

A counselor's responsibilities regarding clients and psychotropic medications include:

- Recognizing when a referral for medication evaluation is appropriate
- Educating clients (and their families, if appropriate) regarding medication as a treatment option

- Making the referral to a physician for medication evaluation when indicated
- Encouraging and monitoring medication compliance
- Inquiring about side effects and target symptoms *at every session*
- Communicating regularly with the physician about the client's status
- Encouraging the client to communicate directly with the physician

When to Refer

As a counselor intern, you should consider referral to a psychiatrist for medication evaluation whenever the client's symptoms are disrupting his or her functioning, or when symptoms have not responded adequately to psychotherapy (Feldman & Feldman, 1997a, 1997b). When in doubt, always seek advice from your site supervisor.

The following list describes clinical situations that always warrant referral for medication consultation. Additional problems certainly necessitate referral to a physician; however, the scope of this chapter is limited to guidelines for referral for medication evaluation.

- There is significant suicidal risk.
- There is risk of injury or accidental death due to self-destructive or risk-taking behaviors.
- There are symptoms of major depression, such as anhedonia (inability to enjoy formerly pleasurable activities); dysphoric mood (*e.g.,* sadness, irritability); sleep disturbance; decreased or increased appetite; low energy; difficulty concentrating; psychomotor retardation; extreme fatigue; feelings of hopelessness, worthlessness, or guilt; and thoughts of death.

  Note: In children, depression may be manifested as an oppositional attitude, angry behavior, preoccupation with somatic complaints, sadness, separation anxiety, expressing a wish to die, or drawing pictures or writing notes about death.

- There are signs of a manic episode, including abnormally elevated or irritable mood, pressured speech, greatly increased energy, psychomotor agitation, reduced sleep, and excessive involvement in activities with a high potential for destructive consequences (*e.g.,* sudden high frequency of unprotected sexual activity with multiple partners, unusual number of costly items purchased, or large amount of money spent in a short time period).

  Note: In children, symptoms of mania may be expressed as irritability, hyperactivity, prolonged rages, and hypersexualized and/or destructive behavior.

- The client presents with symptoms of psychosis, such as hallucinations, delusions, or otherwise impaired reality testing.
- There are either obsessive thoughts or compulsive behaviors, or both.
- The client complains of panic or unremitting anxiety.
- There has been significant sleep deprivation due to insomnia.

- The client has been unable (or unwilling) to eat and has lost significant weight.
- The presenting problem seems to be a recurrence of a past problem that responded favorably to treatment with psychotropic medications.

The above symptoms and situations can be caused by a wide variety of medical illnesses, physical conditions, and chemical exposure, as well as by psychiatric disorders (Morrison, 1997). Therefore, if clients with such symptoms have not had a recent physical examination, they should always be referred back to their physicians to have one, in conjunction with the referral for psychotropic medication.

TALKING TO CLIENTS ABOUT MEDICATION

In combined treatment, the protocol for both pharmacotherapy and psychotherapy sessions should include an inquiry about progress made or problems encountered in the complementary component of the treatment. The agenda for every psychotherapy session should include questions pertaining to medication, including a review of symptoms and questions about compliance and side effects (Feldman & Feldman, 1997a).

Initiating the Discussion About Medication Referral

When initiating a discussion about referral for psychotropic medication, it is important to provide clients with information regarding their disorder and the rationale for taking medication, as well as how the medication can improve functioning (Beck, 2001; Feldman & Feldman, 1997a, 1997b). It is also essential to try to elicit and correct clients' misinformation about medication (Beck, 2001). When the counselor provides sufficient accurate information and explores the client's feelings with respect and empathy, the therapeutic relationship is strengthened and all aspects of treatment are enhanced.

Counselors should explain that both psychological and biological factors have contributed to the client's problem, and that medication will help reduce uncomfortable symptoms whereas psychotherapy will help teach problem-solving skills and new behaviors (Riba & Balon, 2001). Many clients respond well to a statement such as, "The medication can lift your mood and make it easier for you to concentrate in our counseling sessions, so you will be more able to learn to use new ways to cope with stress." The statement should be tailored to the specific needs and situation of the client. For example: "The medication can help reduce your anxiety, so you will be more able to work on important issues in therapy," or "The medication can relieve your racing thoughts, so you will be more able to focus on finding good solutions for your problems."

Delineating Professional Roles

Clients sometimes assume that a referral for medication means the counselor intends to relinquish care to the psychiatrist. Misunderstandings can be avoided by explaining clearly to clients that both professionals will continue to participate in the treatment. For example, the counselor might say, "Dr. Smith will talk with you to decide whether medication might be helpful and if so, she will work with you to choose the best one for you. She will most likely see you several times at first, to make sure you are not having any problems with your medication. But once things are going smoothly, she will need to see you only every few months. Your visits to Dr. Smith will not change the frequency of your meetings with me. I will still continue to see you according to our regular therapy schedule even when you are taking medications."

Encouraging Compliance with Psychotropic Medications

About 60% of mental health clients are *noncompliant*, meaning that they do not take their medications regularly or at all (Ellison, 2000). As discussed earlier, counselors see clients much more frequently than physicians do, and therefore encouraging medication compliance in combined treatment is an important counselor responsibility. Talking with clients and identifying obstacles to compliance is the first step in fulfilling our responsibility. It is often helpful to categorize obstacles as practical, such as forgetting to take pills, or psychological, such as not taking them due to fear of side effects (Beck, 2001). These types of obstacles can usually be overcome by maintaining a strong therapeutic alliance, providing sufficient information about the psychotropic medication, exploring clients' thoughts and feelings thoroughly, and using a problem-solving approach (Beck, 2001; Ellison, 2000). For example, we can encourage the client who forgets to take medication to keep her bottle of pills right near her toothbrush, where it will provide a visual cue. When counseling a client who is fearful of side effects, we can provide accurate, sensible information about managing certain side effects, and we can also remind the client that he has the power to talk with his physician about adjusting the dose or changing medications, should the side effects be too bothersome.

Sometimes a client's noncompliance may be due to irrational thoughts or beliefs about the self, the world, accepting help, doctors, or medications (Beck, 2001). In this case, a cognitive-behavioral approach is useful (Beck, 2001). The counselor can help the client identify the irrational beliefs underlying the noncompliance, challenge them, and formulate more logical thoughts.

General Therapeutic Statements

Most clients in Western cultures feel encouraged when the counselor offers one or more of the following therapeutic statements:

- Starting medication is in some ways like buying a new pair of shoes (or jeans). You often need to try on several different pairs in order to find the one just right for you. Sometimes several different medications must be tried, as well, to find the one just right for you—the one that helps you the most and has the fewest side effects.

- Most medications require two to three weeks to reach therapeutic levels in the blood until they begin to help you feel better.

- Sometimes side effects are unpleasant when a medication is first started, but they often subside or disappear with time. Patience helps.

- Taking a medication does not mean you are crazy or weak. In fact, making good use of all available resources is a sign of very good judgment.

- The purpose of medication is to help you feel more like yourself, at your best. Medication cannot make you "high" or change your personality.

SIGNIFICANCE OF MULTICULTURAL ISSUES AND GENDER

Counselors need to be sensitive to multicultural values and symbolic meanings that may affect combined treatment. Culture shapes clients' feelings about mental illness, acceptance of medication, tolerance of side effects, attitude toward compliance, relationships with professional helpers, and dietary and health habits (Ahmed, 2001; Tseng, 2001). Culture also determines physicians' diagnosing and prescribing practices (Ahmed, 2001; Gonzales, Griffith, & Ruiz, 2001). Counselors should be aware that diversity can affect not only the psychological aspects, but also the medical aspects of pharmacotherapy. A growing body of research indicates that gender, race, and ethnicity differentially affect the absorption, metabolism, and excretion of many medications, which in turn influence medication effectiveness and side effects (Gonzales, Griffith, & Ruiz, 2001; Nadelson & Notman, 2001; Robinson, 2002).

UNDERSTANDING CLIENTS' FEELINGS ABOUT MEDICATION

Although pharmacological intervention emerged from a purely biological model, it encompasses many of the same issues and dynamics as other psychotherapeutic treatment modalities (Gabbard, 2000). Counselors need to attend carefully and empathetically to all their clients' concerns and feelings about medications (Feldman & Feldman, 1997a).

Positive Feelings

Information about the effectiveness of psychotropic medications is easily accessible on the Internet and in the media, so that many clients are well informed and expect their therapists to be equally knowledgeable. Such clients may view a referral for medication positively, as evidence that the counselor is an expert, highly qualified professional.

Clients often experience the referral as a caring or nurturing response from the therapist, or as a genuine acknowledgement of their pain (Gabbard, 2000; Kay, 2001). Many clients feel encouraged to continue treatment when there are two clinicians collaborating to try to be helpful (Feldman & Feldman, 1997b). The prescription itself may provide a calming placebo effect, or may serve as a soothing transitional object, helping the client maintain a sense of connection to the counselor and/or physician (Gabrielli, Fornaro, & Luise, 1997). In fact, some clients may carry their bottle of pills with them as a source of comfort, much as a small child carries around a blanket.

Negative Feelings

Although the clinician who thoughtfully decides to make a referral for medication is trying to act in the best interests of the client, such a referral may elicit negative feelings in the client. First of all, the referral transforms the therapeutic dyad into a therapeutic triad by involving a physician in addition to the counselor and the client. This change alone adds complications that can potentially change or damage the original therapeutic alliance (Kay, 2001; Riba & Balon, 2001). Continued communication between clinicians is essential to insure success in combined treatment arrangements (Feldman & Feldman, 1997a, 1997b).

Some clients referred for medication may feel as though they are being rejected because the counselor dislikes them, considers them uninteresting, or has decided they are very sick or "beyond hope" (Gabbard, 2000; Riba & Balon, 2001). Sometimes clients fear that their families will view medication as proof that they are "crazy" or are the cause of all family problems.

Clients frequently worry that taking medication demonstrates weakness of character. In this case, the clinician can offer the following therapeutic response: "If you have a broken leg, you use a cast for support until the bone heals. Using a cast is not a sign of being weak. It is the right way to treat a broken bone. The medication is like a cast—it provides support until you heal, and it is the right way to treat this problem. Medication is not a sign of being weak."

Clients who are on a maintenance regimen of medication and are currently symptom-free may protest that they do not need medication any longer because they are not having any problems. In this situation, a statement such as the following may be helpful: "People who have diabetes remain healthy as long as they take medicine to control it. When they are taking their medicine, they may even feel and look so healthy that it is easy to forget they have diabetes. But if they stop taking their medicine, they soon become very ill. The same is true for people with bipolar illness (or schizophrenia, recurrent major depression, and so forth). Taking the medicine regularly keeps you healthy."

Clients often forget which symptoms were present before they began taking psychotropic medications, and incorrectly attribute illness-related symptoms to their medication. For example, they may complain that the medicine is keeping them awake at night or making it hard to concentrate, and then insist that they must stop the medication because they cannot tolerate these side effects. Clinical wisdom dictates documenting symptoms throughout treatment and sharing this information with clients who need reassurance that particular symptoms are manifestations of the disorder, rather than side effects of the medication.

All negative feelings need to be explored with sensitivity and respect, and addressed actively as issues about treatment. In combined treatment, medication compliance is a primary treatment objective. The clinician hoping to enhance compliance should never react with impatience to the client's feelings or worries about medications.

Splitting

Splitting is a psychological defense mechanism in which the individual unconsciously creates a mental representation of one person as "all good" and another person as "all bad," rather than being able to view persons realistically and integrate both positive and negative qualities into one mental image (Gabbard, 2000). Combined treatment sometimes provokes defensive splitting by the client, who perceives one clinician positively and the other negatively (Riba & Balon, 2001). For example, the psychotherapist may be viewed as less competent and unable to provide help without assistance from the psychiatrist, whereas the psychiatrist may be viewed as more powerful and skilled because of his or her ability to prescribe medications. Alternatively, the psychotherapist may be regarded as kind and understanding because he or she generally has more time to spend with the client and listens to emotional issues, whereas the psychiatrist may be regarded as uninterested or unkind because he or she sees the client for brief, infrequent appointments and focuses on medications and symptoms.

Clinicians should attend carefully to their own countertransference feelings to avoid unconsciously acting out the client's splitting (Goins, 2001). Such acting out should be suspected if one of the clinicians feels enjoyment at being idealized by the client, becomes annoyed at having to share control of the case, or devalues the role of the other clinician (Riba & Balon, 2001). Riba and Balon (2001) write that a counselor who feels ashamed about not having the special powers of the psychiatrist's medication or a psychiatrist who encourages reliance on medication or prescribes medication for a longer time than is necessary may be unconsciously acting out the client's splitting.

Conflict or competition between treating professionals is experienced by many clients as a traumatic reenactment of childhood conflict between their own parents (Feldman & Feldman, 1997b). Both clinicians should be attuned to the potential for splitting because it is always detrimental to treatment. Each clinician should try to convey to the client respect and appreciation for the role and responsibilities of the other clinician. Frequent communication between clinicians not only discourages splitting but also is essential in formulating treatment goals and enhancing medication compliance (Kay, 2001).

Understanding a Client's Request for Medication

With a plethora of advertisements extolling the benefits of medications in treating such disorders as ADHD, depression, generalized anxiety, and social phobia, clients frequently initiate the request for medication referral, and often the request is appropriate. However, in our "take a pill" society, clients may unrealistically expect psychotropic medication to provide an instant cure for all their problems. In addition, it is important to remember that certain clients may be persuaded by their family members or friends to ask for medication. Some of these people may be trying to help, but others may be acting out their own needs to feel important or to control the client.

Requests for medication commonly have latent meaning or emotional significance. Counselors should listen carefully to try to understand the underlying reason for the request. Sometimes clients' demands for medication represent their characteristic defensive strategy of using substances to numb feelings. At other times, a request for medication expresses the client's yearning for more affection from the counselor, or reflects a developmental need for a concrete representation of the counselor's concern. Often, clients are "testing" the counselor to find out whether the counselor recognizes the extent of their distress or whether the counselor considers their problems to be "serious." Finally, the counselor should be alert to a request for medication as a sign that the client is feeling unsatisfied with therapy. Perhaps he or she is overwhelmed with painful affect, possibly because the pace of psychotherapy is too rapid. Maybe the client is feeling frustrated about not making enough progress in therapy. It behooves the counselor to explore all the client's feeling carefully.

FACTS ABOUT PSYCHOTROPIC MEDICATIONS

The following sections of this chapter offer summaries and tables of the seven major classes of psychotropic medications: antidepressants, antianxiety agents, mood stabilizers, antipsychotics, attention deficit disorder medications, sedative/hypnotics (sleep medications), and cognitive aids. Summaries and a table of herbal remedies are also provided, because counselors are likely to treat clients who are using these. In addition, to make it easy for you to look up information about a specific medication, a comprehensive table including all the psychotropic medications and herbal remedies is presented in Appendix F. This table is alphabetized by trade names (*e.g.,* Prozac) and lists generic names, classes, target symptoms or disorders, side effects, and other important information.

The information presented in the following medication summaries and tables is a synthesis of data taken from a variety of reputable sources, including the *Physicians' Desk Reference* (2002); medication package information inserts; and Web sites of the *Journal of the American Medical Association,* the National Institute of Mental Health, Medline Plus Drug, the National Alliance for the

Mentally Ill, Medscape Psychiatry Specialty Home Page, and WebMD. Online addresses for some of these Web sites are provided below. In some cases, we have cited additional references in the text.

It is important to remember that the fast pace of psychopharmacological research insures that any review of psychotropic medications is likely to be outdated soon after its publication. To fulfill their responsibilities to their clients, counselors must keep current with the rapid developments in the field of psychopharmacology. For the most recent, detailed information on prescription drugs and herbal remedies, you are encouraged to refer to the newest editions of the *Physicians' Desk Reference* (2002) and the *Physician's Desk Reference for Herbal Medicines*, 2nd edition (Gruenwald, 2000), or to other current psychopharmacology texts.

There are also several reputable Internet Web sites that offer reliable, comprehensive, and up-to-date information on psychotropic medications. These Web sites, listed below, are very user-friendly and will allow you to search by a medication's trade or generic name, as well as by a disorder:

- Internet Mental Health (www.mentalhealth.com)
- Medline Plus Drug (www.nlm.nih.gov/medlineplus/druginformation.html)
- Medscape Psychiatry Specialty Home Page (www.medscape.com/psychiatryhome)
- The National Alliance for the Mentally Ill (www.nami.org/illness/)
- The National Institute of Mental Health (www.nimh.nih.gov/practitioners/index.cfm)
- WebMD (www.webmd.com)

Off-Label Use of Medications

Off-label use of a medication refers to prescribing a medication to treat a disorder even though the medication has not received official Food and Drug Administration (FDA) approval as a treatment for the particular disorder. Off-label use may begin when a medication that is approved to treat a disorder does not work effectively and physicians try other medications as alternative treatments, or when a medication unexpectedly proves to be effective in treating a problem different from the one for which it was originally prescribed (Kramer, 2002b). For example, the newer mood stabilizers, such as Lamictal and Neurontin, were originally used to treat seizures. Thus, off-label use of medications occurs concomitantly with the development of new medications as well as with the discovery of new indications for existing medications (Kramer, 2002b).

Alcohol and Illicit Drugs

We chose to discuss the risks of combining psychotropic medications with alcohol and/or illicit drugs in this section, rather than repeating this warning for every medication in all the tables. It is *always* unwise to drink alcohol or use illicit drugs when taking psychotropic medications, and in many cases this combination results in grave or lethal outcomes. Counselors should remind all clients who are taking psychotropic medications to refrain from using alcohol or illicit drugs, and to be sure to tell their doctors if they do use them.

ANTIDEPRESSANTS

Antidepressants are used to reduce the affective, cognitive, and behavioral symptoms associated with depression, including dysphoric mood, sleep and appetite disturbances, anhedonia, low energy and fatigue, feelings of guilt, rumination, social withdrawal, hopelessness, helplessness, and thoughts of death. Certain antidepressants also have been approved to treat anxiety, attention deficit disorders, obsessive-compulsive disorder, panic disorder, posttraumatic stress disorder, premenstrual dysphoric disorder, and social phobia. In addition, antidepressants may sometimes be used in the treatment of the following, without formal FDA approval in all cases: addictions, borderline personality disorder, childhood enuresis, chronic pain, eating disorders, insomnia, migraines, premature ejaculation, and schizoaffective disorder (Pollack, Otto, & Rosenbaum, 1996; Reid, Balis, & Sutton, 1997).

Antidepressants have little or no potential for promoting drug dependency, addiction, or tolerance. Significant concerns in using these medications are their troublesome side effects, potential for serious drug or food interaction that may occur with certain types of antidepressants, and lethality of overdose with some of the older antidepressants.

Treatment with antidepressants for three to six weeks is usually required before full therapeutic effects are seen, and many treatment failures are actually due to insufficient trial of the medication. Antidepressants may be prescribed in combination with other drugs to enhance their therapeutic potency (Schatzberg, Cole, & Battista, 2002). Medications used to augment antidepressants include low doses of lithium, new generation antipsychotics, psychostimulants, and thyroid hormone (Brooks, 2001).

Depression is a recurrent illness, and a biological mechanism known as *kindling* results in structural and biochemical changes in the brain that increase the severity of each subsequent episode and decrease the time between episodes following the initial episode of depression. Furthermore, recurrence occurs in 50% of clients having one prior episode of depression and 80% to 90% for those having two prior episodes (Delgado & Gelenberg, 2001). Therefore, as a rule, clients are treated with antidepressants for a minimum of six months after the first episode, to reduce incidence of relapse or recurrence (Delgado & Gelenberg, 2001). Although there are no formal guidelines for treatment of clients who have suffered multiple depressive episodes, many psychiatrists consider it prudent to maintain these clients on antidepressants indefinitely as a preventive measure (Stahl, 1996).

There are four classes of antidepressants: *specific serotonin reuptake inhibitors (SSRIs), atypicals, tricyclics, and monoamine oxidase inhibitors (MAOIs).* The SSRIs and atypicals are the most recently developed and widely prescribed antidepressants. They are generally associated with fewer adverse side effects and lower risk of lethal overdose than the MAOIs and tricyclics.

Specific Serotonin Reuptake Inhibitors (SSRIs)

Specific serotonin reuptake inhibitors produce a broad spectrum of therapeutic benefits. Their mechanism of action is thought to be blockade of reuptake of serotonin, so that this neurotransmitter remains in the synaptic spaces of the brain for an extended time. All SSRIs share class-specific side effects, with minor variations, depending on the particular medication and the individual client. The most common and most troublesome side effect is decreased libido and sexual dysfunction. Other common side effects include anxiety, dizziness, gastrointestinal upset, and insomnia. Benzodiazepines (discussed later in this chapter) are often prescribed to diminish the anxiety and insomnia associated with SSRIs (Delgado & Gelenberg, 2001). Less common side effects of SSRIs include fatigue, somnolence, sweating, tremor, and unusual or vivid dreams. Risk of lethal overdose is extremely low. Fatal drug interactions between some SSRIs and MAOIs have been reported, so these medications should never be used together. Table 8.1 lists trade and generic names, disorder or target symptoms, side effects, and other information for SSRIs.

Table 8.1 Specific Serotonin Reuptake Inhibitors (SSRIs)

Trade Name (Generic Name)	Disorder or Target Symptoms	Side Effects	Other Facts
Celexa (*Citalopram*)	Major depression	Agitation, agranulocytosis, dry mouth, hypotension, insomnia, low white cell count, sexual somnolence, dysfunction, weight gain.	Dangerous drug interactions with MAOIs.
Lexapro (*Escitalopram*)	Major depression	Agitation, blurred vision, dizziness, drowsiness, nausea, sexual dysfunction, weight changes.	Dangerous drug interactions with MAOIs. Lexapro is the active isomer of Celexa.
Luvox (*Fluvoxamine*)	OCD	Anxiety, appetite decrease, dry mouth, GI upset, insomnia, sedation, sexual dysfunction, tremor.	Dangerous drug interactions may occur with Xanax, Hismanal, Seldane, Halcion. Abrupt withdrawal may result in headache, dizziness, nausea.
Paxil (*Paroxetine*)	Major depression; OCD; panic; anxiety	Constipation, dizziness, dry mouth, GI upset, headaches, sedation, sexual dysfunction, sweating.	Flu-like syndrome may occur upon discontinuation unless dose is tapered gradually.

(continued on next page)

Table 8.1 Specific Serotonin Reuptake Inhibitors (SSRIs) *(continued)*

Trade Name *(Generic Name)*	Disorder or Target Symptoms	Side Effects	Other Facts
Prozac **(same drug** **as Sarafem)** *(Fluoxetine)*	Major depression; OCD; panic; premenstrual dysphoric disorder	Anxiety, GI upset, headaches, insomnia, sedation, sexual dysfunction, tremor, weight changes.	Longest half-life of SSRIs. Dangerous drug interactions with MAOIs; wait five weeks after stopping Prozac before taking MAOIs. No research evidence to support increased suicidality, despite media reports (Stahl, 1996). Once-a-week dose available.
Remeron *(Mirtazapine)*	Major depression	Appetite increase with weight gain, dizziness, elevated cholesterol and triglycerides, sedation, somnolence, sexual dysfunction.	Dangerous dug interactions with MAOIs; wait 14 days after stopping Remeron before taking MAOIs.
Sarafem **(same drug** **as Prozac)** *(Fluoxetine)*	Premenstrual dysphoric disorder	Anxiety, constipation, GI upset, headaches, insomnia, sedation, sexual dysfunction, tremor, weight changes.	Same drug as Prozac. Longest half-life of SSRIs. Dangerous drug interactions with MAOIs; wait five weeks after stopping Sarafem before taking MAOIs.
Zoloft *(Sertraline)*	Major depression; OCD; panic; PTSD; social phobia	Dizziness, dry mouth, GI upset, insomnia, sexual dysfunction, somnolence, sweating, tremor.	Dangerous drug interactions with MAOIs; wait 14 days after stopping Zoloft before taking MAOIs. Abrupt withdrawal may result in headache, dizziness, nausea.

Atypical Antidepressants

Atypical antidepressants influence the reuptake and binding of both serotonin and noradrenaline, and, in some cases, also may affect the action of dopamine. The therapeutic benefits and side effects of atypicals vary considerably, but the compliance rate for this class is high overall because side effects are generally mild. Table 8.2 lists trade and generic names, disorder or target symptoms, side effects, and other information for atypical antidepressants.

Tricyclic Antidepressants

Tricyclic antidepressants affect both serotonin and noradrenaline and are highly effective in treating depression. However, tricyclics have a very high risk of lethal overdose and are associated with problematic side effects, so they have

Table 8.2 Atypical Antidepressants

Trade Name (*Generic Name*)	Disorder or Target Symptoms	Side Effects	Other Facts
Cymbalta (*Duloxetine*)	Major depression	Dry mouth, nausea, somnolence.	Does not cause weight changes. Does not raise blood pressure. Sexual dysfunction (delayed orgasm) in males only.
Desyrel (*Trazodone*)	Major depression; anxiety; insomnia	Dizziness, dry mouth, postural hypotension, sedation.	Risk of sustained erection (priapism) in men, requiring emergency medical intervention.
Effexor (*Venlafaxine*)	Major depression; ADD/ADHD	Appetite changes, constipation, dizziness, dry mouth, GI upset, headache, insomnia, sedation, somnolence, sweating, tachycardia.	Effective treatment for melancholic depression; monitor blood pressure at doses ≥150 mg/day.
Serzone (*Nefazodone*)	Major depression	Dizziness, dry mouth, headache, sedation, somnolence.	Does not cause anxiety or insomnia. Low risk of mania-induction in bipolar. Some risk of liver failure.
Wellbutrin (same drug as Zyban) (*Bupropion*)	Major depression; ADHD	Agitation, anxiety, appetite and weight changes, constipation, insomnia, sweating.	Low incidence of sexual dysfunction. Increased risk of seizure at high doses (>400 mg), with initial rapid increase in dosage, or in clients having bulimia or seizure disorders.
Zyban (same drug as Wellbutrin) (*Bupropion*)	Smoking cessation	Agitation, anxiety, appetite and weight changes, constipation, insomnia, sweating.	Increased risk of seizure at high doses (>400 mg), with initial rapid increase in dosage, or in clients having bulimia or seizure disorders.

largely been replaced by newer antidepressants. When tricyclics are used, psychiatrists often prescribe small amounts at one time to prevent a fatal overdose. Nevertheless, clients sometimes hoard pills until they have accumulated enough for a suicide attempt, and therefore careful monitoring of suicidal ideation and feelings of worthlessness or hopelessness is mandatory.

Class-specific side effects are particularly troublesome with tricyclics (Kaplan & Sadock, 2001). Some of these side effects resolve over time and others may be treated symptomatically—for example, by directing clients to sip water or suck on hard candies to relieve dry mouth, or to increase fiber in the diet to reduce constipation. Table 8.3 lists trade and generic names, disorder or target symptoms, side effects, and other information for tricyclics.

Table 8.3 Tricyclic Antidepressants

Trade Name (Generic Name)	Disorder or Target Symptoms	Side Effects	Other Facts
Anafranil (*Clomipramine*)	OCD; major depression	Blurred vision, constipation, dizziness, drowsiness, dry mouth, EKG abnormalities, GI upset, increased appetite, irregular heartbeat, postural hypotension, sedation, tachycardia, tremor, urinary retention, weakness, weight gain.	
Asendin (*Amoxapine*)	Major depression	Blurred vision, constipation, dizziness, drowsiness, dry mouth, EKG abnormalities, GI upset, increased appetite, irregular heartbeat, postural hypotension, sedation, tachycardia, tremor, urinary retention, weakness, weight gain.	
Elavil (*Amitriptyline*)	Major depression; chronic pain	Blurred vision, constipation, dizziness, drowsiness, dry mouth, EKG abnormalities, GI upset, increased appetite, irregular heartbeat, postural hypotension, sedation, tachycardia, tremor, urinary retention, weakness, weight gain.	
Ludiomil (*Maprotaline*)	Major depression	Blurred vision, dry mouth, constipation, sedation, skin rash, weight gain.	
Norpramin (*Desipramine*)	Major depression; ADHD	Blurred vision, constipation, dizziness, drowsiness, dry mouth, EKG abnormalities, GI upset, increased appetite, irregular heartbeat, postural hypotension, sedation, tachycardia, tremor, urinary retention, weakness, weight gain.	
Pamelor (*Nortriptyline*)	Major depression; ADHD	Blurred vision, constipation, dizziness, drowsiness, dry mouth, EKG abnormalities, GI upset, increased appetite, irregular heartbeat, postural hypotension, sedation, tachycardia, tremor, urinary retention, weakness, weight gain.	
Sinequan (*Doxepin*)	Major depression	Blurred vision, constipation, dizziness, drowsiness, dry mouth, EKG abnormalities, GI upset, increased appetite, irregular heartbeat, postural hypotension, sedation, tachycardia, tremor, urinary retention, weakness, weight gain.	

(continued on next page)

Table 8.3 Tricyclic Antidepressants *(continued)*

Trade Name (Generic Name)	Disorder or Target Symptoms	Side Effects	Other Facts
Surmontil (*Trimipramine*)	Major depression	Blurred vision, constipation, dizziness, drowsiness, dry mouth, EKG abnormalities, GI upset, increased appetite, irregular heartbeat, postural hypotension, sedation, tachycardia, tremor, urinary retention, weakness, weight gain.	Toxic in overdose.
Tofranil (*Imipramine*)	Major depression; bedwetting in children	Blurred vision, constipation, dizziness, drowsiness, dry mouth, EKG abnormalities, GI upset, increased appetite, irregular heartbeat, postural hypotension, sedation, tachycardia, tremor, urinary retention, weakness, weight gain.	Increased risk of falls in elderly.
Vivactil (*Protriptyline*)	Major depression	Blurred vision, constipation, dizziness, drowsiness, dry mouth, EKG abnormalities, GI upset, increased appetite, irregular heartbeat, postural hypotension, sedation, tachycardia, tremor, urinary retention, weakness, weight gain.	

Monoamine Oxidase Inhibitors (MAOIs)

Monoamine oxidase inhibitors are especially effective in the treatment of atypical depression. However, risk of lethal overdose is exceedingly high, side effects are problematic, and MAOIs may cause severe, often fatal interactions with a wide range of foods and drugs, so they are not widely used.

Individuals taking MAOIs must avoid various types of cheeses, fruits, meats, and fish, as well as sauerkraut, soy sauce, yeast, and alcoholic beverages. They also must avoid many prescription and over-the-counter drugs, including but not limited to antihistamines, appetite suppressants, asthma medications, cold medications, cough suppressants, local anesthetics such as Novocain, and psychostimulants used to treat ADD/ADHD, such as Ritalin. MAOIs should not be prescribed for clients whose depressive symptoms include memory or concentration problems because remembering which foods and medications to avoid may be difficult or confusing.

Clients using MAOIs should carry a Medic Alert card informing medical personnel about MAOI use (Kaplan & Sadock, 2001). Table 8.4 lists trade and generic names, disorder or target symptoms, side effects, and other information for MAOIs.

Table 8.4 Monoamine Oxidase Inhibitors (MAOIs)

Trade Name (Generic Name)	Disorder or Target Symptoms	Side Effects	Other Facts
Nardil (*Phenelzine*)	Major depression	Blurred vision, dizziness, dry mouth, edema, excitement, GI upset, hypomania, hypertension, hypertensive crisis, postural hypotension, insomnia, urinary hesitancy or retention, sedation, sexual dysfunction, tachycardia, tremor, weight gain.	Dangerous interactions with SSRIs and many other Rx and OTC drugs and foods. Abrupt discontinuation may cause agitation, anxiety, nightmares, psychosis. Hypertensive crisis is acute medical emergency.
Parnate (*Tranylcypromine*)	Major depression	Dizziness, dry mouth, constipation, excitement, GI upset, headaches, hypertension, hypertensive crisis, hypomania, insomnia, postural hypotension, sedation, tachycardia.	Dangerous interactions with SSRIs and many other Rx and OTC drugs and foods. Abrupt discontinuation may cause agitation, anxiety, confusion, depression, delirium, headaches, insomnia, muscle weakness, nightmares, psychosis. Hypertensive crisis is acute medical emergency.

ANTIANXIETY AGENTS

Anxiety is a normal, adaptive human reaction to threat that evolved as a component of the fight-or-flight survival response (Stahl, 1996). However, anxiety is considered maladaptive when it impedes functioning or is aroused by persons or situations that do not pose any real danger. *Antianxiety agents* (also known as *anxiolytics*) are used to alleviate the uncomfortable physical and psychological symptoms that may be caused by anxiety, including chills, clammy hands, difficulty concentrating, difficulty swallowing, dizziness, dry mouth, fatigue, fearfulness, frequent urination, gastrointestinal upset, headaches, insomnia, irritability, muscle tension and aches, sweating, tachycardia, and tremor.

Antianxiety agents are used in the treatment of acute stress reactions, adjustment disorder with anxious mood, agitated depression, agoraphobia, alcohol withdrawal, elective mutism, generalized anxiety disorder, insomnia, night terrors, panic disorder, performance anxiety, phobias, posttraumatic stress, obsessive-compulsive disorder, and rage attacks (Kaplan & Sadock, 2001; Pollack, Otto, & Rosenbaum, 1996; Reid, Balis, & Sutton, 1997). Antianxiety agents are also used as muscle relaxants, to control seizures, as pre-anesthetic agents for surgery, to treat migraines, and as short-term treatment for some of the side effects associated with SSRIs and atypical antidepressants.

There are five classes of antianxiety agents: barbiturates, azaspirone, benzo-diazepines, antihistamines, and antihypertensives. Benzodiazepines are the most widely used antianxiety medications today. Barbiturates comprise the oldest class, but are rarely used anymore because they are highly addictive, very se-dating, have a high potential for lethal overdose, and may cause severe, life-threatening withdrawal symptoms when they are discontinued. The only bar-biturate used with some frequency is phenobarbital, and it seldom is prescribed for anxiety. (Note: If you would like further information concerning phe-nobarbital, you can find it in Appendix F.)

Azaspirone

Buspar is the only medication in the azaspirone class. It is the most recently in-troduced antianxiety medication and provides safe and effective treatment for anxiety. Buspar is less sedating than the other antianxiety agents and does not impair coordination or balance. It does not interact adversely with alcohol, nor does it promote addiction, dependency, or tolerance. However, treatment for one to three weeks is required before clients realize a therapeutic effect from Buspar, and this is frequently too long to wait for individuals suffering acute anxiety.

Table 8.5 lists trade and generic names, disorder or target symptoms, side effects, and other information for Buspar.

Table 8.5 Azaspirone

Trade Name (Generic Name)	Disorder or Target Symptoms	Side Effects	Other Facts
Buspar (*Buspirone*)	Anxiety	Agitation, dizziness, drowsiness, headaches, GI upset, insomnia.	Three-week lag before effects are felt. Ineffective for panic disorder or very severe anxiety. Adverse interaction with MAOIs. No risk for addiction.

Benzodiazepines

Benzodiazepines have less risk of lethal overdose and fewer side effects than the barbiturates, and are therefore widely prescribed for the treatment of anxiety. Approximately 10% of the U.S. population uses benzodiazepines each year (Kaplan & Sadock, 2001). Several of these medications have been developed specifically for the treatment of insomnia and are known as hypnotics, hypno-anxiolytics, or sedative/hypnotics.

Discontinuation of benzodiazepines may be associated with severe and sometimes life-threatening withdrawal symptoms, including abdominal cramps, convulsions, fever, muscle aches, sweating, tremor, and vomiting (Kaplan & Sadock, 2001). Thus, abrupt withdrawal should be avoided and doses should be tapered when discontinuing these medications.

Benzodiazepines are associated with several other problems. First, they produce a *rebound effect*, which is a temporary recurrence and worsening of pretreatment symptoms (Kaplan & Sadock, 2001). The rebound effect is especially troublesome when the initial target symptom was insomnia, because the insomnia returns with increased intensity when the medication is discontinued. Second, class-specific side effects of benzodiazepines include excessive sedation, drowsiness, and difficulty concentrating, which can cause increased risk of injury from falls, especially in elderly clients (Kaplan & Sadock, 2001). Third, benzodiazepines have been associated with memory loss, but fortunately this tends to resolve when the drug is discontinued.

Finally, benzodiazepines have a moderate potential for producing dependency, addiction, and tolerance. Therefore, the lowest possible therapeutic dose should be prescribed, treatment duration should not be prolonged, and clients with a history of substance abuse should be monitored carefully.

Table 8.6 lists trade and generic names, disorder or target symptoms, side effects, and other information for benzodiazepines.

Table 8.6 Benzodiazepines

Trade Name (Generic Name)	Disorder or Target Symptoms	Side Effects	Other Facts
Ativan (*Lorazepam*)	Anxiety	Dizziness, drowsiness, fatigue, headache, muscle cramps, nausea, vomiting, weakness.	Some potential for drug tolerance and dependence.
Dalmane (*Flurazepam*)	Anxiety; insomnia	Dizziness, drowsiness, fatigue, headache, muscle cramps, nausea, vomiting, weakness.	Adverse interaction with alcohol. Some risk for addiction.
Klonopin (*Clonazepan*)	Bipolar; aggression; panic; anxiety; insomnia; Tourette's; restless legs syndrome; aggressive behavior in children	Dizziness, drowsiness, fatigue, headache, muscle cramps, nausea, vomiting, weakness.	
Librium (*Chlordiaz-epoxide*)	Anxiety	Dizziness, drowsiness, fatigue, headache, muscle cramps, nausea, vomiting, weakness.	Slow onset. Some risk for addiction.
Restoril (*Tenazepam*)	Anxiety; insomnia	Dizziness, drowsiness, fatigue, headache, muscle cramps, nausea, vomiting, weakness.	Some risk for addiction.
Serax (*Oxazepam*)	Anxiety	Dizziness, drowsiness, fatigue, headache, muscle cramps, nausea, vomiting, weakness.	Slow onset. Some risk for addiction.
Tranxene (*Clorazepate*)	Anxiety	Dizziness, headache, skin rash.	Moderate risk for addiction.

(continued on next page)

Table 8.6 Benzodiazepines *(continued)*

Trade Name (Generic Name)	Disorder or Target Symptoms	Side Effects	Other Facts
Valium (*Diazepam*)	Anxiety	Dizziness, drowsiness, fatigue, headache, muscle cramps, nausea, vomiting, weakness.	Rapid action. High risk for addiction.
Xanax (*Alprazolam*)	Anxiety; panic	Dizziness, drowsiness, fatigue, headache, muscle cramps, nausea, vomiting, weakness.	Some risk for addiction.

Antihistamines

Although their primary use is in the treatment of allergic reactions, *antihistamines* are sometimes prescribed to reduce symptoms of anxiety, to treat insomnia, or to alleviate the side effects of antipsychotic medications. Antihistamines are highly sedating, have few side effects, and do not promote addiction, dependency, or tolerance.

Table 8.7 lists trade and generic names, disorder or target symptoms, side effects, and other information for antihistamines.

Table 8.7 Antihistamines

Trade Name (Generic Name)	Disorder or Target Symptoms	Side Effects	Other Facts
Benadryl (*Diphenhydramine*)	Insomnia; treatment of side effects of antipsychotics	Dizziness, dry mouth, drowsiness, GI upset, itching, sedation, skin rash, sweating, thickening of respiratory secretions, urinary retention, wheezing.	Contraindicated in individuals with asthma, peptic ulcers, prostate problems, urinary problems.
Vistaril (*Hydroxyzine*)	Anxiety	Drowsiness, dry mouth, sedation, tremor.	Adverse interactions with alcohol, analgesics, barbiturates, narcotics.

Antihypertensives

Antihypertensives were developed to treat hypertension (high blood pressure) and cardiac arrhythmias, but sometimes are prescribed to reduce the symptoms of anxiety. They are very effective in alleviating the physical symptoms of anxiety, such as rapid heartbeat or sweating, but are not quite as effective in reducing the psychological symptoms, such as a feeling of impending doom or difficulty concentrating (Kaplan & Sadock, 2001). In addition, they sometimes can precipitate a depressive episode. Table 8.8 lists trade and generic names, disorder or target symptoms, side effects, and other information for antihypertensives.

Table 8.8 Antihypertensives

Trade Name (Generic Name)	Disorder or Target Symptoms	Side Effects	Other Facts
Catapres (*Clonidine*)	Anxiety; aggressive behavior and ADHD in children	Dizziness, dry mouth, constipation, fatigue, sedation.	Primary use is treatment for hypertension. Abrupt withdrawal may cause agitation, anxiety, headache, hypertension.
Inderal (*Propanolol*)	Anxiety	Depression, fatigue, hypotension, sedation, tachycardia.	Contraindicated for individuals with asthma or diabetes.
Tenex (*Guanfacine*)	Aggressive behavior and ADHD in children	Constipation, dizziness, drowsiness, dry mouth.	
Tenormin (*Atenolol*)	Anxiety	Depression, dry eyes, headache, sexual dysfunction, sedation, skin rash, sore throat.	Contraindicated for individuals with kidney disease. Abrupt withdrawal may cause cardiac problems.

MOOD STABILIZERS
(ANTICONVULSANTS AND LITHIUM)

Mood stabilizers were the first class of psychotropic medications used both to treat mental illness and to prevent its recurrence (Stahl, 1996). Anticonvulsants and lithium are used, either separately or combined, to alleviate the symptoms of mania and hypomania, and to stabilize the abnormal mood fluctuations characteristic of bipolar disorder (Kaplan & Sadock, 2001). Symptoms of mania and the manic phase of bipolar disorder may include decreased need for sleep, psychomotor agitation, racing thoughts, pressured speech, grandiosity, involvement in potentially self-destructive activities (*e.g.*, indiscriminate shopping sprees), and impaired thinking.

Other indications for mood stabilizers include treatment of aggression, alcohol withdrawal, atypical psychosis, depression with psychotic features, hallucinations associated with chronic substance abuse, mania, panic attacks, schizoaffective disorders, schizophrenia, and seizures (Gabbard, 2001; Kaplan & Sadock, 2001).

Mood stabilizers are associated with potentially serious conditions, so that clients taking these medications must be regularly monitored through blood tests. Levels of lithium that exceed the narrow therapeutic margin in the blood can be toxic and possibly lethal. Furthermore, because lithium is excreted directly through the kidneys, clients must be careful to maintain a steady water volume in the body by drinking plenty of fluids and avoiding exercise-induced dehydration. Anticonvulsants may be associated with serious or lethal blood and liver disorders, and clients must be vigilant about early warning signs, such as easy bruising, bleeding, fever, sores in the mouth, and abnormal blood count. Table 8.9 lists trade and generic names, disorder or target symptoms, side effects, and other information for mood stabilizers.

Table 8.9 Mood Stabilizers

Trade Name (Generic Name)	Disorder or Target Symptoms	Side Effects	Other Facts
Depakote (*Valproic acid*)	Bipolar; migraine	Appetite decrease, blood disorders, GI upset, itching, menstrual changes, tardive dyskinesia.	Potentially fatal liver damage may occur. Local irritation of mouth may occur if tablets are chewed before swallowing.
Eskalith (**same drug as Lithobid**) (*Lithium carbonate*)	Bipolar	Acne, ataxia, confusion, drowsiness, dry mouth, GI upset, hair loss, hallucinations, memory problems, thyroid problems, tremor, urinary problems, weight gain.	Narrow margin of safety; toxic or lethal if too high; monitor blood levels, thyroid function, kidney function.
Lamictal (*Lamatrogine*)	Bipolar	Ataxia, blurred vision, cough, dizziness, nausea and vomiting, rhinitis, skin rash, sore throat, somnolence.	Risk of accidental injury due to poor coordination and dizziness. Must discontinue if skin rash develops.
Lithobid (**same drug as Eskalith**) (*Lithium carbonate*)	Bipolar	Acne, ataxia, confusion, drowsiness, dry mouth, GI upset, hair loss, hallucinations, memory problems, thyroid problems, tremor, urinary problems, weight gain.	Narrow margin of safety; toxic or lethal if too high; monitor blood levels; regulate liquid intake and output.
Neurontin (*Gabapentin*)	Bipolar; anxiety; neuropathic pain	Ataxia, dizziness, double vision, fatigue, rhinitis, somnolence.	Take at bedtime to decrease side effects.
Tegretol (*Carbamazepine*)	Bipolar	Ataxia, blood disorders, blurred vision, constipation, dizziness, dry mouth, drowsiness.	May precipitate mania. Adverse interaction with MAOIs. Risk for rare but lethal skin and blood conditions. Monitor with blood tests.
Topamax (*Topiramate*)	Bipolar	Anxiety, breast pain in women, constipation, dizziness, drowsiness, fatigue, GI upset, nausea, vision problems; weight loss.	May cause change in sense of taste. Often prescribed concurrently with antipsychotics to prevent weight gain.
Trileptal (*Oxcarbazepine*)	Bipolar; chronic pain	Dizziness, drowsiness, hyponatremia.	Does not require monitoring of blood levels or liver function. May reduce effectiveness of birth control pills.

ANTIPSYCHOTICS

Antipsychotics are used to reduce the symptoms of psychosis, a condition in which the client is unable to differentiate what is real from what is not real. These symptoms may include cognitive symptoms, such as delusions, hallucinations, or any otherwise severely impaired reality testing, and behavioral manifestations, such as aggression, agitation, belligerence, catatonia, or disorganized speech. Psychosis may be present in schizophrenia, schizoaffective disorder, depression with psychotic features, bipolar disorder, substance abuse, and transiently in borderline personality disorder. Other indications for antipsychotics include anxiety, autism, debilitating nausea, delusional disorder, delirium, dementia, schizotypal personality disorder, and Tourette's syndrome. Antipsychotics are also highly effective when combined with antidepressants in the treatment of mood disorders with psychotic features (Kaplan & Sadock, 2001; Pollack, Otto, & Rosenbaum, 1996; Reid, Balis, & Sutton, 1997).

There are two groups of antipsychotics: *traditional* and *new generation*. The *traditional* antipsychotics alleviate the positive symptoms of psychosis, such as agitation, delusions, hallucinations, disorganized thinking and speech, and bizarre behavior. However, these medications are ineffective in treating the negative symptoms, such as attention impairment, blunted affect, difficulty in abstract thinking, emotional and social withdrawal, lack of goal-directed behavior, limited motivation, passivity, and poverty of speech (Stahl, 1996). In addition, side effects, such as significant weight gain, are severe and cause many persons to discontinue their medications.

In the past few years, *new generation* antipsychotics have been developed that are effective in targeting not only the positive symptoms, but also the negative symptoms, thus enhancing interpersonal relationships and reducing the sense of isolation suffered by clients with psychotic disorders. These new generation medications are associated with fewer and less severe side effects than the traditional antipsychotics. Two of the newest antipsychotics, Abilify and Geodon, do not cause weight gain.

All antipsychotics are associated with unpleasant side effects that cause many clients to stop taking their medication, although the side effects are less severe with the new generation antipsychotics. Some of the more adverse class-specific side effects include *acute dystonic reaction, akathisia, akinesia, tardive dyskinesia*, and a life-threatening condition known as *malignant neuroleptic syndrome*. These terms are explained below:

- *Acute dystonic reaction* involves severe muscle spasm and stiffness, frequently in the neck, face, or back. The head and neck may be pulled to one side sharply by strong muscle contractions. Acute dystonic reaction occurs in 10% of individuals taking antipsychotic medications, usually within a week of beginning treatment, and responds well to medical treatment.

- *Akathisia* refers to a feeling of intense motor restlessness, relieved only by pacing, swinging the legs, or jumping up and down, and is a common side effect of antipsychotics. It occurs in 40% of clients after a single dose of antipsychotic medication and in as many as 75% of clients who take these medications for one week (Gitlin, 1996).

■ *Akinesia* is a condition that occurs weeks or months after initiation of treatment with antipsychotics, and mimics the symptoms of Parkinson's disease. It involves a mask-like expression, shuffling gate, drooling, hand tremor, stiff joints, and slow movement (Gitlin, 1996).

■ *Tardive dyskinesia* is an irreversible syndrome that is associated with the traditional antipsychotic medications. It involves involuntary, repetitive muscular movements, such as grimacing, sticking out the tongue, smacking the lips, blinking, turning the neck and head, and twisting the extremities. Tardive dyskinesia is permanent and untreatable, and is correlated with dose size and duration of treatment. Therefore, psychiatrists strive to prescribe the lowest effective dose of antipsychotics, for the shortest amount of time, to prevent this syndrome from occurring. At the present time, Clozaril is the only antipsychotic shown not to be associated with tardive dyskinesia. There is not enough long-term research to determine whether the other new generation antipsychotics lead to tardive dyskinesia.

■ *Malignant neuroleptic syndrome* is a rare but potentially life-threatening reaction characterized by coma or confusion, blood pressure fluctuations, kidney failure, very high fever, muscle rigidity, rapid heartbeat, and respiratory distress. This condition requires emergency medical treatment. Malignant neuroleptic syndrome is most likely to occur upon initiation of treatment with antipsychotics, following rapid increase in dose, or after prolonged treatment at a high dose.

Table 8.10 lists trade and generic names, disorder or target symptoms, side effects, and other information for the traditional antipsychotics and Table 8.11 provides the same information for the new generation antipsychotics.

Table 8.10 Traditional Antipsychotics

Trade Name (Generic Name)	Disorder or Target Symptoms	Side Effects	Other Facts
Haldol (*Haloperidol*)	Schizophrenia; psychotic features of mood disorder; Tourette's	Agitation, akathisia, akinesia, anxiety, depression, dizziness, dry mouth, excess salivation, fatigue, insomnia, nasal congestion, rigidity, sedation, tardive dyskinesia, tremor, weight changes.	Associated with tardive dyskinesia.
Loxitane (*Loxapine*)	Schizophrenia; psychotic features of mood disorder	Akathisia, blurred vision, dizziness, dry mouth, fatigue, hair loss, headache, hypotension, nasal congestion, pseudo-Parkinsonism, tachycardia, tardive dyskinesia, tremor, urinary retention.	Contraindicated in patients with epilepsy. Associated with tardive dyskinesia.

(continued on next page)

Table 8.10 Traditional Antipsychotics *(continued)*

Trade Name *(Generic Name)*	Disorder or Target Symptoms	Side Effects	Other Facts
Mellaril *(Thioridazine)*	Schizophrenia; psychotic features of mood disorder	Blurred vision, cardiac arrhythmias, constipation, dizziness, drowsiness, dry mouth, hypotension, nasal congestion, sexual dysfunction, skin rash, swelling, tardive dyskinesia, urinary retention, weight gain.	Adverse interactions with anesthetics, barbiturates, narcotics. Avoid insecticides. Associated with tardive dyskinesia.
Navane *(Thiothixene)*	Schizophrenia; psychotic features of mood disorder	Akathisia, dizziness, drowsiness, dry mouth, hypotension, nasal congestion, pseudo-Parkinsonism, skin rash, tardive dyskinesia, urinary retention, weight gain.	Adverse interactions with anesthetics, barbiturates, narcotics. Associated with tardive dyskinesia.
Prolixin *(Fluphenazine)*	Schizophrenia; psychotic features of mood disorder	Agitation, blurred vision, drowsiness, dry mouth, excess salivation, GI upset, insomnia, sexual dysfunction, tardive dyskinesia, weight gain.	Adverse interactions with alcohol, epinephrine, hypnotics. Contraindicated in patients with liver disease. Associated with tardive dyskinesia.
Stelazine *(Trifluoperazine)*	Schizophrenia; psychotic features of mood disorder	Akathisia, appetite decrease, cardiac arrest, constipation, dizziness, drowsiness, hair loss, headache, muscle weakness, pseudo-Parkinsonism, skin rash, tremor, tardive dyskinesia.	Adverse interactions with alcohol, anesthetics, antianxiety agents, barbiturates, narcotics. Associated with tardive dyskinesia.
Thorazine *(Chlorpromazine)*	Schizophrenia; psychotic features of mood disorder	Akinesia, blurred vision, constipation, depression, dry mouth, drowsiness, EKG abnormalities, menstrual changes, muscle weakness, pseudo-Parkinsonism, skin pigmentation, tachycardia, tardive dyskinesia, weight gain.	Adverse interactions with alcohol, anesthetics, antianxiety agents, anticoagulants, anticonvulsants, barbiturates, epinephrine, narcotics. Associated with tardive dyskinesia.

Table 8.11 New Generation Antipsychotics

Trade Name (Generic Name)	Disorder or Target Symptoms	Side Effects	Other Facts
Abilify (*Aripiprazole*)	Schizophrenia	Agitation, headache, insomnia.	Newest antipsychotic developed and causes fewest side effects. Unique mechanism of action: dopamine system stabilizer.
Clozaril (*Clozapine*)	Schizophrenia; psychotic features of mood disorder; suicidal ideation	Anxiety, constipation, convulsions, depression, dizziness, drowsiness, dry mouth, excess salivation, GI upset, headache, shortness of breath, skin rash, tachycardia, weight gain.	Contraindicated in patients with cardio-vascular, eye, kidney, liver, prostate, or urinary problems. Potentially fatal blood disease may occur. Not associated with tardive dyskinesia.
Geodon (*Ziprasidone*)	Schizophrenia; psychotic features of mood disorder	Abnormal muscle move-ments, constipation, cough, diarrhea, dizzi-ness, fatigue, restless-ness, rhinitis, skin rash, tremor.	Only antipsychotic that is associated with little or no weight gain. Controls both positive and negative symptoms of psychosis. Risk of tardive dyskinesia not yet known.
Risperdal (*Risperidone*)	Schizophrenia; psychotic features of mood disorder; borderline personality	Constipation, dizziness, drowsiness, weight gain.	Controls both positive and negative symptoms. Avoid alcohol. Risk of tardive dyskinesia not yet known.
Seroquel (*Quetiapine*)	Schizophrenia; psychotic features of mood disorder; borderline personality	Dizziness, drowsiness, dry mouth, headache, tardive dyskinesia.	Controls both positive and negative symptoms. Avoid alcohol. Risk of tardive dyskinesia not yet known.
Trilafon (*Perphenazine*)	Schizophrenia; psychotic features of mood disorder	Akathisia, blurred vision, constipation, dizziness, dry mouth, nasal congestion, pseudo-Parkinsonism, somnolence, tardive dyskinesia, tremor, weight gain.	Contraindicated in patients with depression, kidney problems, or liver problems. Adverse interactions with alcohol, antihistamines, analgesics, barbiturates, opiates. Associated with tardive dyskinesia.
Zyprexa (*Olanzapine*)	Schizophrenia; psychotic features of mood disorder; borderline personality	Akathisia, constipation, dizziness, postural hypotension, somnolence, weight gain.	Controls both positive and negative symptoms. Risk of tardive dyskinesia not yet known.

ATTENTION DEFICIT
DISORDER MEDICATIONS
(PSYCHOSTIMULANTS AND
NORADRENALINE REUPTAKE INHIBITORS)

Several classes of medications are used to treat the symptoms of attention deficit disorders (ADD and ADHD). These include *psychostimulants*, *antidepressants*, and a recently developed *noradrenaline reuptake inhibitor (NRI)* known as Strattera. (Because antidepressants are described in an earlier part of this chapter, only the psychostimulants and Strattera will be discussed in this section.) Attention deficit disorder medications are used to decrease motor activity, problems concentrating, impulsivity, forgetfulness, emotional lability, memory problems, and distractibility.

Psychostimulants were originally developed to aid in weight loss, but are widely used to manage the symptoms of ADD and ADHD in children and adults (Korn, 2001). Approximately 75% of persons with ADD or ADHD will benefit from psychostimulants (Hampton, 2001; Simon, Etkin, Godine, Heller, Kuter, Shellito, & Stern, 1998). Although the mechanism of action of psychostimulants is not well understood, it is thought that they alter the reuptake and binding of noradrenaline (Korn, 2001). Other indications for psychostimulants include treatment of apathy in the elderly; chronic fatigue syndrome; narcolepsy; and treatment-resistant depression associated with Alzheimer's disease, dementia, and AIDS (Kaplan & Sadock, 2001; Pollack, Otto, & Rosenbaum, 1996; Reid, Balis, & Sutton, 1997; Stahl, 1996). Class-specific side effects of psychostimulants include insomnia, appetite suppression, and temporary growth suppression in children (Kaplan & Sadock, 2001). Exacerbation or precipitation of tics and Tourette's syndrome may occur with the use of psychostimulants.

There is little risk of addiction to the psychostimulants used to treat ADD and ADHD, because they are taken orally and blood levels rise slowly (Simon et al., 1998). However, the pills produce a "high" when crushed and inhaled, and 16% of children and adolescents who take Ritalin report pressure from peers who want to buy it from them (Simon et al., 1998). Counselors need to be cognizant of the fact that children and adolescents may sell their psychostimulants to their schoolmates, and that parents sometimes may take their children's medications. We have found it helpful to use an open-ended question when inquiring about a client's compliance with psychostimulants, such as, "How have things been going with regard to the Ritalin?" Often, this type of questioning allows clients and their families to talk more openly than they would if the counselor asked, "Have you (or your child) been taking the Ritalin exactly the way your doctor prescribed?" One mother was able to tell us honestly that she was taking the medication herself, because her child was so out of control that she needed it more than he did. We then could empathize with the mother's stress, provide information about the effects of the medication, offer guidelines on parenting a child with ADHD, and help her obtain a referral for an appointment for herself.

In 2003, a new, nonstimulant medication, Strattera, was approved for the treatment of ADD and ADHD. The mechanism of action of Strattera is through the inhibition of noradrenaline reuptake. This medication is associated with fewer side effects and a more stable therapeutic blood level than the psychostimulants. Another advantage is that Strattera is not a controlled substance, and therefore physicians may call in prescriptions by phone and may make multiple refills available for patients.

Table 8.12 lists trade and generic names, disorder or target symptoms, side effects, and other information for psychostimulants, and Table 8.13 lists the same information for the NRI Strattera.

Table 8.12 Psychostimulants

Trade Name (Generic Name)	Disorder or Target Symptoms	Side Effects	Other Facts
Adderall (same drug as Concerta, Metadate, and Ritalin) (*Methylphenidate hydrochloride*)	ADD/ADHD	Appetite suppression, dizziness, dry mouth, GI upset, headache, hypertension, impotence, insomnia, irritability, palpitations, tachycardia, tremor, weight loss.	Associated with psychological dependence.
Concerta (same drug as Adderall, Metadate and Ritalin) (*Methylphenidate hydrochloride*)	ADD/ADHD	Appetite suppression, depression, dizziness, growth suppression, GI upset, insomnia, weight loss.	Once-a-day dose. Associated with psychological dependence.
Cylert (*Pemoline*)	ADD/ADHD	Appetite suppression, growth suppression, insomnia, weight loss.	Risk of liver damage.
Metadate (same drug as Adderall, Concerta, and Ritalin) (*Methylphenidate hydrochloride*)	ADD/ADHD	Appetite decrease, headache, insomnia.	Once-a-day dose. Associated with psychological dependence.
Provigil (*Modafinil*)	Narcolepsy	Anxiety, appetite suppression, headache, infection, insomnia, nausea.	Used off-label to treat ADD/ADHD and Alzheimer's disease.
Ritalin (same drug as Adderall, Concerta, and Metadate) (*Methylphenidate hydrochloride*)	ADD/ADHD	Appetite suppression, depression, dizziness, growth suppression, GI upset, insomnia, weight loss.	Associated with psychological dependence and drug tolerance.

Table 8.13 Noradrenaline Reuptake Inhibitor

Trade Name (Generic Name)	Disorder or Target Symptoms	Side Effects	Other Facts
Strattera (*Atomoxetine*)	ADD/ADHD	Dry mouth, GI upset, insomnia, weight loss.	Newest FDA-approved medication for ADD/ADHD.

SEDATIVE/HYPNOTICS
(SLEEP MEDICATIONS)

Sleep disturbance is a problem that may be associated with everyday stress, as well as with almost all emotional and mental disorders. Because sleep deprivation and fatigue exacerbate symptoms and impair functioning, physicians often prescribe medications to help improve sleep, in conjunction with other psychotropic medications. Table 8.14 lists trade and generic names, disorder or target symptoms, side effects, and other information for sedative/hypnotics (sleep medications).

Table 8.14 Sedative/Hypnotics (Sleep Medications)

Trade Name (Generic Name)	Disorder or Target Symptoms	Side Effects	Other Facts
Ambien (*Zolpidem*)	Insomnia	Agitation, amnesia, dizziness, drowsiness, headache, nausea, nightmares.	Short-acting. May cause rebound effect. Some potential for abuse.
Halcion (*Triazelam*)	Insomnia	Dizziness, drowsiness, fatigue, headache, muscle cramps, nausea, vomiting, weakness.	
Imovane (*Zopiclone*)	Insomnia	Agitation, amnesia, dizziness, drowsiness, headache, nausea, nightmares.	Short-acting. May cause rebound effect. Some potential for abuse.
Sonata (*Zaleplon*)	Insomnia	Agitation, amnesia, dizziness, drowsiness, headache, nausea, nightmares.	Short-acting. May cause rebound effect.

COGNITIVE AIDS

Many persons with Alzheimer's disease are treated in a mental health setting; therefore, counselors may encounter clients who are taking medications known as *cognitive aids*. These medications may improve memory and may slow

Table 8.15 Cognitive Aids

Trade Name (Generic Name)	Disorder or Target Symptoms	Side Effects	Other Facts
Aricept (*Donepezil hydrochloride*)	Alzheimer's disease	Anorexia, diarrhea, fatigue, insomnia, muscle cramps, nausea, vomiting.	Slows progress of disease in some patients, but does not cure Alzheimer's disease.
Exelon (*Revastigmine*)	Alzheimer's disease	Abdominal pain, dizziness, headache, nausea, somnolence, tremor, vomiting, weakness, weight loss.	Slows progress of disease in some patients, but does not cure Alzheimer's disease.
Reminyl (*Galantamine HBr*)	Alzheimer's disease	Appetite decrease with weight loss, diarrhea, dizziness, GI upset, muscle weakness, sweating, vomiting.	Slows progress of disease in some patients, but does not cure Alzheimer's disease.

down the progressive cognitive decline associated with Alzheimer's, but they do not cure the disease. Table 8.15 lists trade and generic names, disorder or target symptoms, side effects, and other information for cognitive aids.

HERBAL REMEDIES

Although research findings are mixed, a few studies indicate that herbal remedies may be useful in treating mild depression, anxiety, and insomnia. However, herbs are not effective for treatment if these problems are moderate to severe (Hypericum Depression Trial Study Group, 2002). Some clients may be more willing to try herbal remedies because they believe that these drugs are less stigmatizing, safer, or more "natural" than prescription medications. Yet it is important to remember that many potent prescription medications were derived from herbs and other plants. "Natural" and "herbal" do not necessarily mean "safe" and "risk-free." Herbs may be toxic and can cause serious side effects (Brooks, 2001; Miller, 1998). In fact, a warning issued by the FDA in 2002 cautioned that one popular herbal remedy (Kava Kava) can cause serious liver damage (Warner, 2002). Herbal remedies also can interfere with the effectiveness of certain prescription medications, and can cause dangerous interactions with others (Brooks, 2001; Miller, 1998).

Herbal remedies are not subject to federal regulation, so the amount and quality of active ingredients may vary greatly by manufacturer and even by each bottle (Brooks, 2001). Persons taking herbal remedies should always inform their physicians. Table 8.16 lists trade and generic names, disorder or target symptoms, side effects, and other information for herbal remedies.

Table 8.16 Herbal Remedies

Trade Name (Generic Name)	Disorder or Target Symptoms	Side Effects	Other Facts
Kava Kava (*Piper methysticum*)	Anxiety; insomnia	Somnolence; long-term use (more than three months) may result in skin rash, blood abnormalities, facial swelling, muscle weakness.	Increases side effects of antianxiety drugs, antidepressants, sleep aids, pain relievers, muscle relaxants. May result in coma when used with alprazolam. New FDA warning about potentially severe liver damage.
St. John's Wort (*Hypericum*)	Mild to moderate major depression; anxiety	Restlessness.	Lowers effectiveness of birth control pills and drugs used to treat HIV (protease inhibitors and nonnucleoside reverse transcriptase inhibitors). May increase side effects of SSRIs and MAOIs. Increases risk of sunburn.
Valerian Root (*Valeriana officinalis*)	Anxiety; insomnia; premenstrual dysphoric disorder	Somnolence, dizziness. Long-term use may cause headaches, excitability, restlessness, insomnia, tachycardia.	Increases side effects of muscle relaxants, sleep aids, antianxiety drugs, pain relievers, antidepressants.

CONCLUDING REMARKS

Due to the increasing availability of safe and effective psychotropic medications, the mental health field is at risk of becoming skewed toward a purely medical or biological approach to the treatment of mental illness (Gabbard, 2001). Pressures from third-party payers to provide "quick fix" treatment for medical and psychological problems exacerbates this distorted perspective. However, clinicians should keep in mind that such an approach is reductionist, because the etiology and course of mental and emotional disorders always involve a dynamic interplay of biological, psychological, and social factors (Gabbard, 2001). Psychotherapy or counseling is an integral part of treatment, even for disorders known to have a largely biological basis, such as schizophrenia or bipolar disorder.

Psychotropic medications alleviate a wide spectrum of psychological symptoms, but they must be prescribed and monitored with great care, and the potential risks must be weighed judiciously against the potential benefits for every client on an individual basis. In addition, although these medications improve mood, cognition, and physical symptoms, psychotherapy is needed to help people develop skills to manage the psychosocial stressors associated with their problems (Schatzberg, Cole, & Battista, 2002). Optimal treatment entails a holistic approach, with attention not only to the biological underpinnings of mental and emotional disorders, but to all aspects of the person. As is true in all counseling situations, counselors and counselor interns treating clients who are taking psychotropic medications should strive to use a comprehensive, biopsychosocial vantage point, to address all realms of the client's life experience.

REFERENCES

AHMED, I. (2001). Psychological aspects of giving and receiving medications. In W. S. Tseng and J. Streltzer (Eds.), *Culture and psychotherapy* (pp. 123–134). Washington, DC: American Psychiatric Publishing.

BECK, J. S. (2001). A cognitive approach to medication compliance. In J. Kay (Ed.), *Integrated treatment of psychiatric disorders* (pp. 113–140). Washington, DC: American Psychiatric Publishing.

BROOKS, J. (2001). *What are the drugs used for depression? A Well-Connected Report: Depression*. Retrieved on October 5, 2001, from the WebMD Web site: my.webmd.com/content/article/1680.51008

CASEY, B. J., GIEDD, J. N., & THOMAS, K. M. (2000). Structural and functional brain development and its relation to cognitive development. *Biological Psychiatry, 54*, 241–257.

COYLE, J. T. (2000). Psychotropic drug use on very young children. *Journal of the American Medical Association, 283*, 1059.

DELGADO, P. L., & GELENBERG, A. J. (2001). Antidepressant and antimanic medications. In G. O. Gabbard (Ed.), *Treatment of psychiatric disorders* (3rd ed., Vol. 2, pp. 1131–1179). Washington, DC: American Psychiatric Publishing.

ELLISON, J. M. (2000). Enhancing adherence in the pharmacotherapy treatment relationship. In A. Tasman, M. Riba, & K. Silk (Eds.), *The doctor-patient relationship in pharmacotherapy* (pp. 71–94). New York: Guilford.

EL-MALLAKH, R. S., PETERS, C., & WALTRIP, C. (2000). Antidepressant treatment and neural plasticity. *Journal of Child and Adolescent Psychopharmacology, 10*(4), 287–294.

EPSTEIN, H. T. (2001). An outline of the role of brain in human cognitive development. *Brain and Cognition, 45*, 44–51.

FELDMAN, L. B., & FELDMAN, S. L. (1997a). Conclusion: Principles for integrating psychotherapy and pharmacotherapy. *In Session: Psychotherapy in Practice, 3*(2), 99–102.

FELDMAN, L. B., & FELDMAN, S. L. (1997b). Introduction. *In Session: Psychotherapy in Practice, 3*(2), 1–3.

GABBARD, G. O. (2000). *Psychodynamic psychiatry in clinical practice* (3rd ed.). Washington, DC: American Psychiatric Publishing.

GABBARD, G. O. (2001). Mind and brain in psychiatric treatment. In G. O. Gabbard (Ed.), *Treatment of psychiatric disorders* (3rd ed., Vol. 2, pp. 3–19). Washington, DC: American Psychiatric Publishing.

GABRIELLI, F., FORNARO, P., & LUISE, L. (1997). Psychopharmacological therapy and its effects as transitional object. *Biological Psychiatry, 42*(1), (Supplement 1), 172S.

GITLIN, M. J. (1996). *The psychotherapist's guide to psychopharmacology* (2nd ed.). Toronto: The Free Press, Macmillan.

GOINS, M. K. (2001). Split treatment: The psychotherapy role of the prescribing psychiatrist. *Psychiatric Services, 52*(5), 605–609.

GONZALES, C. A., GRIFFITH, E. E. H., & RUIZ, P. (2001). Cross-cultural issues in psychiatric treatment. In G. O. Gabbard (Ed.), *Treatment of psychiatric disorders* (3rd ed.,Vol. 1, pp. 47–67).Washington, DC: American Psychiatric Publishing.

GRUENWALD, J. (Ed.). (2000). *Physician's desk reference for herbal medicines.* Montvale, NJ: Medical Economics Company.

HAMPTON,T.W. (2001). *What are the medications for attention-deficit hyperactivity disorder? A Well-Connected Report: Attention-deficit hyperactivity disorder.* Retrieved on October 5, 2001, from the WebMD Web site: my.webmd. com/viewarticle/1680.52950

HERLENIUS, E., & LAGERCRANTZ, H. (2001). Neurotransmitters and neuromodulators in early human development. *Early Human Development, 65*(1), 21–37.

HOLLON, S. D., & FAWCETT, J. (2001). Combined medication and psychotherapy. In G. O. Gabbard (Ed.), *Treatment of psychiatric disorders* (3rd ed.,Vol. 2, pp. 1247–1266).Washington, DC: American Psychiatric Publishing.

HYPERICUM DEPRESSION TRIAL STUDY GROUP. (2002). Effect of *Hypericum perforatum* (St. John's Wort) in major depressive disorder. *Journal of the American Medical Association, 287,* 1807–1814.

IACONO, R. P. (2000). *Overview of neurotransmitters and chemical synapses.* Retrieved on February 14, 2002, from the Neuroscience Clinic Web site: www.pallidotomy.com/ neurotransmitters.html

INGERSOLL, R. E. (2001).The nonmedical therapist's role in pharmacological intervention with adults. In E. R.Welfel & R. E. Ingersoll (Eds.), *The mental health desk reference* (pp. 88–93). New York: John Wiley & Sons.

KANE, J. M., & MALHOTRA, A. K. (2001). Clinical psychopharmacology of schizophrenia and psychotic disorders. In G. O. Gabbard (Ed.), *Treatment of psychiatric disorders* (3rd ed.,Vol. 1, pp. 1027–1043).Washington, DC: American Psychiatric Publishing.

KAPLAN, H. I., & SADOCK, B. J. (2001). *Kaplan & Sadock's pocket handbook of clinical psychiatry* (3rd ed.). Baltimore: Williams and Wilkins.

KAY, J. (Ed.). (2001). *Integrated treatment of psychiatric disorders.*Washington, DC: American Psychiatric Publishing.

KORN, M. L. (2001). *Understanding and treating adult ADHD. US Psychiatric and Mental Health Congress 2001.* Retrieved on April 5, 2002, from the Medscape Web site: www.medscape. com/view article/412883

KRAMER,T. A. M. (2002a). Dopamine system stabilizers. *Medscape Psychiatry and Mental Health eJournal, 7*(1). Retrieved on March 30, 2002, from the Medscape Web site: www.medscape. com/viewarticle/429444

KRAMER,T. A. M. (2002b). A field of increasing possibilities. *Medscape General Medicine 4*(4). Retrieved on December 17, 2002, from the Medscape Web site: www.medscape.com/ viewarticle/446142

LYDIARD, B. R., OTTO, M. O., & MILROD, B. (2001). Panic disorder. In G. O. Gabbard (Ed.), *Treatment of psychiatric disorders* (3rd ed.,Vol. 2, pp. 1447–1483).Washington, DC: American Psychiatric Publishing.

MARTIN, A., KAUFMAN, J., & CHARNEY, F. D. (2000). Pharmacotherapy of early-onset depression: Update and new directions. *Child and Adolescent Pediatric Clinics of North America, 9*(1), 135–157.

MILLER, L. G. (1998). Herbal medicines: Selected clinical considerations focusing on known or potential drug-herb interactions. *Archives of Internal Medicine, 158,* 2200–2211.

MORRISON, J. (1997). *When psychological problems mask medical disorders.* New York: Guilford.

NADELSON, C. C., & NOTMAN, M. T. (2001). Gender issues in psychiatric treatment. In G. O. Gabbard (Ed.),

Treatment of psychiatric disorders (3rd ed., Vol. 1, pp. 21–45). Washington, DC: American Psychiatric Publishing.

NATIONAL INSTITUTE FOR HEALTH CARE MANAGEMENT. (2002). *Prescription drug expenditures in 2001: Another year of escalating costs.* Retrieved on March 30, 2002, from the National Institute for Health Care Management Web site: www.nihcm. org/spending.2001.pdf

NATIONAL INSTITUTE OF MENTAL HEALTH. (2000). *Treatment of children with mental disorders. NIH Publication No. 00–4702.* Retrieved on March 3, 2002, from the National Institute of Mental Health Web site: www.nimh.nih.gov/publicat/childqa.cfm

NATIONAL INSTITUTE OF MENTAL HEALTH. (2001). *The teenage brain: A work in progress. NIH Publication No. 01–4929.* Retrieved on March 10, 2002, from the National Institute of Mental Health Web site: www.nimh.nih.gov/publicat/teenbrain.cfm

NEMEROFF, C. B., SCHATZBERG, A. F., GOLDSTEIN, D. J., DETKE, M. J., MALLINCKRODT, C., LU, Y., & TRAN, P.V. (2002). Duloxetine for treatment of major depressive disorder. *Psychopharmacology Bulletin, 36*(4), 106–132.

NEUROTRANSMITTERS. (2002). Retrieved on February 4, 2002, from Salmon, the University of Plymouth Department of Psychology Web site: salmon.psy.plym.ac.uk/year1/neurotr.htm

PHYSICIANS' DESK REFERENCE. (2002). Montvale, NJ: Medical Economics Company.

POLLACK, M. H., OTTO, M. W., & ROSENBAUM, J. F. (1996). *Challenges in clinical practice: Pharmaceutical and psychosocial strategies.* New York: Guilford.

REID, W. H., BALIS, G. U., & SUTTON, B. J. (1997). *The treatment of psychiatric disorders* (3rd ed.) (Revised for DSM-IV). Bristol, PA: Brunner/Mazel.

RIBA, M. B., & BALON, R. (2001). The challenges of split treatment. In J. Kay (Ed.), *Integrated treatment of psychiatric disorders* (pp. 143–164). Washington, DC: American Psychiatric Publishing.

ROBINSON, G. E. (2002). *Women and psychopharmacology. Medscape Women's Health eJournal, 7*(1). Retrieved on March 10, 2002, from the Medscape Web site: medscape.com/viewarticle/423938

RUSHTON, J. L., & WHITMIRE, J. T. (2001). Pediatric stimulant and selective serotonin reuptake inhibitor prescription trends. *Archives of Pediatric and Adolescent Medicine, 155*, 560–565.

SCHATZBERG, A., COLE, J., & BATTISTA, C. (2002). *Manual of clinical pharmacology* (4th ed.). Washington, DC: American Psychiatric Publishing.

SCHATZBERG, A. F., & NEMEROFF, C. B. (2001). *Essentials of clinical psychopharmacology.* Washington, DC: American Psychiatric Publishing.

SIMON, H., ETKIN, M. J., GODINE, J. E., HELLER, D., KUTER, I., SHELLITO, P. C., & STERN, T. A. (1998). *What are the general guidelines for treating attention-deficit hyperactivity disorder? A Well-Connected Report: Attention-deficit hyperactivity disorder in children.* Retrieved on October 5, 2001, from the WebMD Web site: my.webmd.com/printing/dmk/dmk_article_5461048

STAHL, S. M. (1996). *Essential psychopharmacology: Neuroscientific basis and practical applications.* Cambridge, United Kingdom: Cambridge University Press.

TSENG, W. S. (2001). Culture and psychotherapy: An overview. In W. S. Tseng & J. Streltzer (Eds.), *Culture and psychotherapy.* Washington, DC: American Psychiatric Publishing.

WARNER, J. (2002). *Kava may cause liver damage. MedscapeWire.* Retrieved on April 6, 2002, from Medscape Web site: www.medscape.com/viewarticle/430762

WEBSTER'S UNABRIDGED DICTIO-NARY. (2001). New York: Random House.

ZITO, J. M., SAFER, D. J., dosREIS, S., GARDNER, J. F., BOLES, M., & LYNCH, F. (2000). Trends in the prescribing of psychotropic medications to preschoolers. *Journal of the American Medical Association, 283,* 1025–1030.

ZITO, J. M., SAFER, D. J., dosREIS, S., MAGDER, L. S., GARDNER, J. F., & ZARIN, D. A. (1999). Psychotherapeutic medication patterns for youths with attention-deficit–hyperactivity disorder. *Archives of Pediatric and Adolescent Medicine, 153,* 1257–1263.

9

Professional Challenges

D uring your internship, you will most likely be faced with a variety of challenging experiences that are inherent in the counseling profession. This chapter presents practical, basic guidelines for managing some of the anxiety-provoking situations and professional dilemmas that you may encounter for the first time as a counselor intern. We do not attempt to outline explicit treatment methods for specific clinical problems or disorders, but rather address situations that frequently leave counselor interns feeling unsettled and wondering how to proceed. For information on ways to treat a particular clinical condition (*e.g.*, panic attacks), you should access other resources.

We strongly encourage you to use additional resources whenever you have questions concerning your interactions with clients. Discuss complex situations or dilemmas with your supervisor or other clinicians at your internship site; ask your counseling program professors for help; read professional books or journal articles that provide current clinical information; refer to the *Code of Ethics and Standards of Practice* of the American Counseling Association; and review policies and procedures outlined by your agency.

The particular needs, problems, and goals of your client, as well as the demands and constraints of the immediate clinical circumstances, will dictate your treatment strategies, goals, and therapeutic interventions. The guiding principle in every case, however, involves working toward engagement of the client in a strong therapeutic alliance. This therapeutic alliance enables you to collect data for assessing the client, develop a compassionate and evolving understanding of the client, formulate an appropriate treatment plan, and begin helping (Shea, 1998). We have identified the following general goals that provide a common platform for the engagement process and are well-suited for most counseling situations:

- Provide a safe space, physically and psychologically
- Establish rapport
- Employ a collaborative stance
- Understand your client's concerns contextually
- Instill hope
- Identify and mobilize your client's internal and external resources and support systems
- Identify and mobilize your own resources and support systems

The following question and answer section lists common clinical challenges in alphabetical order, and provides recommendations for managing some of the perplexing, complicated, or stressful treatment situations you may experience during your counseling internship. We have also provided suggested readings wherever possible, so that you may explore certain areas in more depth.

ABUSE AND NEGLECT

Reporting It

How should I report suspected abuse or neglect of a child, adolescent, disabled person, or older adult?

Professional helpers, such as physicians, nurses, teachers, social workers, and counselors, are mandated in all 50 states to report suspected abuse of children under 18 years old, disabled persons, and older adults (Kalichman, 2001). Reporting abuse or neglect is a difficult task, because the counselor is likely to be distressed at hearing of the situation, and clients often are upset by the need to report it. The perpetrator may be a family member whom the abused or neglected individual loves and does not want to "get in trouble." Other family members may be very upset and angry not only with the perpetrator, but also with the victim and the counselor. Many counselor interns describe their own experience in making an abuse report as "traumatic." If you find, during your internship, that you must report an incident of abuse or neglect, use your site supervisor, your counseling program supervisor, and other professors from your counseling department for support. Sharing your feelings with other students in your internship discussion group may also be helpful.

As soon as you begin your internship placement, familiarize yourself with your state laws pertaining to reporting abuse and neglect, and collect the telephone numbers of the agencies that must be notified. Call the agency and ask what usually happens when they receive a report, so that you will be able to share this information with the client, should the situation occur.

Consult with your site supervisor and counseling program supervisor to discuss what protocol to follow in case you suspect or learn that abuse or neglect is occurring. Your agency may have specific guidelines concerning how to report. You should follow all agency, county, and state procedures carefully, and document everything in detail. When a report is made, agencies usually ask for the following information:

- The name, age, and address of the individual being abused or neglected
- In the case of a child victim, the name and address of the child's parent or guardian
- The name, age, and address of the perpetrator (if known)
- The exact nature of the abuse or neglect
- The location and date where the abuse or neglect occurred
- Your name, position, and relationship to the victim, perpetrator, or informant

Always make certain, during your initial discussion with clients concerning the limits of confidentiality, to discuss the fact that you are required by state law to report all cases of suspected abuse or neglect (Kalichman, 1999). Then, if clients should disclose abuse during their sessions, they will not feel betrayed by a sudden breach of confidentiality when you tell them you must make such a report. If your client asks you what will happen after the abuse or neglect is reported, present the facts gently and with sensitivity, but be honest about all procedures. Often, a social worker will come to interview a child at school or will visit the home. Sometimes the individual who was abused or neglected will need a physical examination. The perpetrator may be charged with criminal activity. A child or older adult may be removed from the home. If the abused or neglected person is an older adult and is mentally competent, he or she should be an active participant in identifying the problem and deciding on the most reasonable plan of action (Danzinger, 2001).

We have found it most helpful if you can encourage clients to make the telephone call to report abuse or neglect, while you sit nearby and offer support. If they are unwilling or unable, it is a good idea for you to make the call yourself immediately, in the client's presence. In the case of a child, you should request that the parent make the call or be present when you make the call, if at all possible. Reporting abuse or neglect in the client's presence models being honest, taking action to keep people safe, and behaving in accordance with the law. Document in detail everything said during the phone call, as well as the fact that the client was present. We have found that these practices prevent later misrepresentations or accusations about what was reported. Early in our career, one of us called to report child abuse without the parent being present. The county social worker who subsequently visited the home misstated the facts that we had reported, causing unnecessary rupture of our counseling relationship with the client.

Suggested Readings on Abuse and Neglect

BARNETT, O., MILLER-PERRIN, C., & PERRIN, R. (1997). *Family violence across the lifespan.* Thousand Oaks, CA: Sage.

GELLES, R. (1997). *Intimate violence in families.* Thousand Oaks, CA: Sage.

GIARDINO, A. P., & GIARDINO, E. R. (Eds.). (2002). *Recognition of child abuse for the mandated reporter.* St. Louis, MO: G. W. Medical Publications.

HERMAN, J. (1997). *Trauma and recovery* (2nd ed.). New York: Basic Books.

KALICHMAN, S. (1999). *Mandated reporting of suspected child abuse: Ethics, law, and policy* (2nd ed.). Washington, DC: American Psychological Association.

MALCHIODI, C. (1997). *Breaking the silence: Art therapy with children from* *violent homes* (2nd ed.). New York: Brunner/Mazel.

MEYERS, J. E. B., BERLINER, L., BRIERE, J., & REID, T. (Eds.). (2002). *The APSAC handbook on child maltreatment* (2nd ed.). Thousand Oaks, CA: Sage.

TOWER, C. C. (2002). *Understanding child abuse and neglect.* Boston: Allyn & Bacon.

WINTON, M. A., & WINTON, B. A. (2001). *Child abuse and neglect: Multi-disciplinary approaches.* Boston: Allyn & Bacon.

WOLFE, D., McMAHON, R., & PETERS, R. (Eds.). (1997). *Child abuse: New directions in prevention and treatment across the lifespan.* Thousand Oaks, CA: Sage.

ADOLESCENTS

Interacting Therapeutically with Adolescents

I will be seeing an adolescent client during my internship, and his parents have informed me that he has told them he doesn't want to come for counseling and won't talk during his sessions. What approach would be most helpful when I counsel adolescents?

Adolescence, according to Erik Erikson (1963), is characterized by a search for identity and the consolidation of disparate elements of the self into an integrated whole that is congruent with self-perceptions and evaluation by others. Adolescents are faced with some very difficult developmental tasks as they struggle to find their way from childhood to adulthood. Corey and Corey (1997) describe adolescence as a turbulent and possibly lonely period, characterized by many conflicts and paradoxes. During adolescence, boys and girls usually:

- Relinquish their dependency on the adults in their lives
- Begin to separate from their families
- Begin to make decisions that will affect their futures
- Begin to relate to the opposite sex in new ways
- Learn to interact with their peers successfully so that they will belong to a group
- Must contend with powerful media and peer-group messages urging them to experiment with alcohol, drugs, and sex

- Undergo the rapid physical development and emotional changes of puberty

In addition, adolescents must face the pressures and stressors of today's fast-paced, increasingly complex, and decreasingly family-oriented society. The result is often teenage depression, eating disorders, alcohol and drug abuse, sexual promiscuity and pregnancy, and juvenile delinquency (Dacey & Travers, 1999; Turner & Helms, 1995). The suicide rate for teenagers has increased more than 300% since 1960; 13 adolescents kill themselves each day, and there are about 300,000 teen suicide attempts each year, with guns or poison being the most common means used (Turner & Helms, 1995).

When you counsel adolescent clients, you will need to work hard to establish enough trust to build a therapeutic relationship. You may have more success if you are honest, reliable, and consistent, and if you demonstrate your caring, support, interest, and respect (Hanna, Hanna, & Keys, 1999).

Group therapy is often the most effective treatment modality for adolescents because of their strong allegiance to their peers (Corey & Corey, 2002; Dulcan & Martini, 1998). The therapy group provides a forum for teenagers to find that they are not alone with their problems, to experience a sense of belonging, to experiment with interpersonal interactions with their peers and the adult group leader(s), to learn to accept and provide honest feedback, and to explore values.

Suggested Readings on Adolescents

BERMAN, A. L., & JACOBS, D. A. (1996). *Adolescent suicide: Assessment and intervention.* Washington, DC: American Psychological Association.

COMMITTEE ON ADOLESCENCE. (1996). *Adolescent suicide.* Washington, DC: American Psychiatric Press.

COREY, G., & COREY, M. (2002). *Groups: Process and practice* (6th ed.). Pacific Grove, CA: Brooks/Cole.

GIL, E. (1996). *Treating abused adolescents.* New York: Guilford.

KESSLER, J. (1996). *Psychopathology of childhood.* Englewood Cliffs, NJ: Prentice-Hall.

O'NEILL, R., HORNER, R., ALBIN, R., SPRAGUE, J., STOREY, K., & NEWTON, J. (1997). *Functional assessment and program development for problem behavior.* Pacific Grove, CA: Brooks/Cole.

CHILDREN

Interacting Therapeutically with Children

One of my clients is a six-year-old child who has behavioral problems. How can I help this youngster?

Counseling children often requires that you modify your professional role, your style, and your treatment modalities. You need to be cognizant of the influence of various "systems" on your client, such as the family, the school, the neighborhood group of children, and sometimes the court and/or other social service agencies. Most of the time, a child client is brought to counseling because one or more adults believe there is a problem, not because the child believes there is a problem. Children are dependent on the adults around them, and therefore you may find that you need to act as an advocate for your client and as a liaison to all these powerful outside forces.

Counseling children is more complex than counseling adults, because you are faced with a triad rather than a dyad in your treatment relationship. Therefore, you will need to build a therapeutic relationship not only with the child, but also with the child's primary caregiver. In addition, you must strive to encourage a healthy relationship between the child and his or her caregiver.

If you work with children during your internship, you need to be very familiar with developmental theory, so that you can assess your client's cognitive, emotional, social, behavioral, and physical development. A thorough understanding of where the child stands developmentally allows you to:

- Relate to the child most effectively to establish trust and a therapeutic alliance
- Determine whether problems or symptoms represent significant deviations from normal development
- Identify goals for change
- Formulate treatment plans in specific areas, to help children "get back on track" developmentally

Young children often express themselves more easily through art and play therapy than through the more traditional "talking therapy." Even very verbal, bright children may not possess the vocabulary or linguistic ability to describe their perceptions, affective experiences, attitudes, or difficulties (Thompson & Rudolph, 2000). Children can more easily "play out" or draw powerful feelings, upsetting thoughts, worrisome situations, or past events. Play or art are helpful treatment modalities because they:

- Engage the child immediately by providing an enjoyable, nonthreatening activity and attractive materials
- Afford the child emotional release (catharsis)
- Separate painful or troubling images from the child's self, so they are not so secret anymore
- Increase the bond between yourself and the child as you share these feelings and experiences
- Empower the child by giving him or her some symbolic control over situations where, in reality, he or she has little or no control

For play therapy, we recommend simple toys, such as a family of dolls, stuffed animals, hand puppets, blocks, cars and trucks, and perhaps a tea set, doctor set, or basic board game. Art materials should also be simple and geared to the child's developmental level, as well as to your own tolerance for messiness. You may select crayons, markers, Play-Doh, collage materials, or various types of paint.

To understand your child client's play and art, you will need to remember to think metaphorically (Dulcan & Martini, 1998). The play or art is a symbolic representation of the child's inner self and his or her perceptions of the world. We recommend that you keep your interpretations and interventions within the metaphor that the child has created. Do not connect your interventions too closely to the child, because the child will feel threatened and communication will be hampered. For example, it is helpful to say, "Tell me about the sad face you drew on this person," instead of, "Do you feel sad like the person you drew in your picture?"

Cognitive behavioral therapy is an effective treatment for depression and anxiety in older children (Kearney & Linning, 2001; Reinecke, Ryan, & DuBois, 1998). Depressed and anxious children can be taught to monitor their thoughts, identify cognitive distortions, and substitute more realistic thoughts to reduce their uncomfortable feelings (McWhirter & Burrow, 2001). In addition to this type of *cognitive restructuring*, counselors may use psychoeducation, relaxation training, systematic desensitization, and role play when working with older children (Kearney & Linning, 2001).

Behavioral therapy is appropriate for many childhood problems and for almost any age child, and is especially useful when counseling children with ADHD (Barkley, 1998). Behavioral strategies for counseling children typically include establishing small, clear, concrete behavioral goals within a limited time frame and providing positive reinforcement when these goals are achieved. For example, one goal might be for the child to remember to bring all homework assignments home from school for three consecutive days. The child would be rewarded with a sticker each day that he or she brought homework assignments home and would be given a small treat or privilege when the three-day goal had been accomplished. Goals can be increased and time periods can be gradually extended as the child achieves success. Parents and teachers need to be closely involved, so that the child experiences consistent support and guidance in developing self-control (Barkley, 1998).

Working successfully with a child client involves being creative and flexible in tailoring your therapeutic approach to your client's developmental level, as well as being sensitive to the child's unique personality and needs. Patience, warmth, a sense of humor, and an honest enjoyment of children are also helpful characteristics of the child therapist.

Suggested Readings on Children

BROMFIELD, R. (1997). *Playing for real: Exploring the world of child therapy and the inner worlds of children.* Northvale, NJ: Jason Aronson.

DULCAN, M., & MARTINI, D. D. (1998). *Concise guide to child and adoles-* *cent psychiatry* (2nd ed.). Washington, DC: American Psychiatric Press.

MALCHIODI, C. (1997). *Breaking the silence: Art therapy with children from violent homes* (2nd ed.). New York: Brunner/Mazel.

McWHIRTER, J., McWHIRTER, B., McWHIRTER, A., & McWHIRTER, E. (1998). *At-risk youth: A comprehensive response* (2nd ed.). Pacific Grove, CA: Brooks/Cole.

ORTON, G. (1997). *Strategies for counseling with children and their parents.* Pacific Grove, CA: Brooks/Cole.

THOMPSON, C., & RUDOLPH, L. (2000). *Counseling children* (5th ed.). Pacific Grove, CA: Brooks/Cole.

TOWER, C. C. (2002). *Understanding child abuse and neglect.* Boston: Allyn & Bacon.

WINTON, M. A., & WINTON, B. A. (2001). *Child abuse and neglect: Multi-disciplinary approaches.* Boston: Allyn & Bacon.

CLIENTS WITH CHRONIC ILLNESS AND/OR DISABILITIES

Understanding the Special Needs of Your Clients

How can I be most helpful to clients who are coping with chronic illness or disabilities?

Having a physical or mental disability or a chronic medical illness can be associated with significant psychological distress (Brems, 2001). Furthermore, a wide range of medical conditions can cause depression, anxiety, or even psychosis (Morrison, 1997). Psychological disorders or emotional upset often occur concomitantly with physical and mental challenges and with chronic illnesses, and more than half of patients seeing their primary care physician for a medical illness also suffer from a diagnosable emotional disorder (Wickramasekera, Davies, & Davies, 1996). You should relate to your clients who are coping with chronic illness or disability (for example, AIDS, spinal cord injury, cancer, chronic obstructive pulmonary disease, or mental retardation) in the same way you relate to all your clients: by demonstrating your empathy, your interest, your respect, your nonjudgmental acceptance, your genuineness, your caring, your trustworthiness, and your desire to help. Your counseling goals, however, will be determined by the severity, scope, and chronicity of your clients' medical or physical status (Brems, 2001). In addition, you will need to be aware of and monitor your client's condition and to communicate with the other healthcare providers who are involved with your client (Brems, 2000, 2001).

You will need to be creative and flexible and to ask your client what he or she finds most helpful. You should learn as much as possible about the unique needs and problems associated with your client's special situation. You can do research on the Internet or at the library, and contact organizations that offer specialized information (*e.g.,* the Muscular Dystrophy Association, the American Cancer Association, and the National Institutes of Health). We suggest also that you consult with your site supervisor and your counseling program supervisor, who may be able to share their insights and experiences with you.

Suggested Readings on Clients with
Chronic Illness and/or Disabilities

BREMS, C. (2000). *Dealing with challenges in psychotherapy and counseling.* Pacific Grove, CA: Brooks/Cole.

DROTAR, D. (2000). *Promoting adherence to medical treatment in chronic childhood illness: Concepts, methods, and interventions.* New York: Lawrence Erlbaum.

FALVO, D. R. (1999). *Medical and psychosocial aspects of chronic illness and disability.* New York: Aspen.

JONGSMA, A. E., & SLAGGERT, K. (2000). *The mental retardation and*

developmental disability treatment planner. New York: John Wiley & Sons.

KASCHAK, E. (Ed.). (2001). *Minding the body: Psychotherapy in cases of chronic and life-threatening illness.* New York: Haworth Press.

MARSHAK, L. E., SELIGMAN, M., & PREZANR, F. (1999). *Disability and the life cycle.* New York: Basic Books.

OLKIN, R. (2001). *What psychotherapists should know about disability.* New York: Guilford Press.

CONFIDENTIALITY

Discussing the Limits of Confidentiality

How can I explain the limits of confidentiality to my adult clients and still expect that they will trust me enough to disclose personal information? How can I explain confidentiality to children in a way they will be able to understand?

Legal considerations, as well as our own personal values and professional ethical standards, often dictate the parameters of our counseling relationship (Corey, Corey, & Callahan, 1998). The sanctity of the counselor–client alliance and professional confidentiality are no longer absolute legal standards. Therefore, you must always disclose the limits of confidentiality to your clients at the outset of your counseling relationship. Corey (2001) writes that, in general, confidentiality is breached when clients pose a danger to themselves or others; when the counselor suspects that a child under age 18, a disabled person, or an older adult is being abused or neglected; when a client requires hospitalization; when clients sign forms for records to be released; and when a court action demands information from the client's record. Find out your exact state regulations concerning the situations in which you will need to break confidentiality, and memorize these.

As a novice counselor, you may find the task of defining confidentiality to your clients awkward or threatening. However, most clients will be reassured to have this information, because many of them will be wondering about the privacy of what they talk about during their counseling sessions. We recommend that you try writing out your explanation of confidentiality several different ways, and then practice saying it, by yourself, until you feel most comfortable. When talking to clients, we stress that strict confidentiality is maintained most of the time, and that confidentiality is of primary importance to us.

We explain confidentiality to our adult clients this way:"Everything we say to each other during our counseling sessions is private, except in certain instances. By law, I am required to get help for you if you feel like hurting yourself or need to be hospitalized for other reasons, and I am required to report to the proper authorities if you tell me that you have plans to hurt another person or if I learn of any abuse of children, disabled persons, or older adults. In addition, in some cases, a court may request information from your records. In all of these situations, I would discuss my plans with you first, before acting. My only motive would be to protect you and others from harm, to help keep people safe. I'd be glad to answer any questions or hear any comment you may have about what I've just said."

During your internship, and as long as you require supervision, you will also need to discuss your status as a counselor trainee. You may wish to say, "The state requires that all mental health professionals be supervised by more experienced helpers for a number of years to ensure the highest quality of care for all clients. I am currently being supervised by Dr. Smith, who is a Professional Clinical Counselor [or whatever title the supervisor holds]. I will be reviewing your records with him from time to time, so that he can offer suggestions concerning our counseling sessions. However, Dr. Smith will keep everything absolutely confidential."

Finally, if you are working with a third-party payer that requests information detailing the client's diagnosis and treatment, you should inform the client which facts will be disclosed. We have found it helpful to discuss with clients exactly what we will write on treatment plans or requests for authorization for sessions from third-party payers. Some clients opt to pay out of pocket, rather than use their insurance benefits, in order to protect their privacy.

For children, we modify our explanation of confidentiality a bit:"The things we talk about in here are mostly private, just between us, because it's important that you can tell me anything and everything you are feeling or thinking. But it's really important for me to keep you and other people safe, too. So if you tell me something that makes me worry that you might hurt yourself or somebody else, or if I hear that a child or older person is being hurt, then I would have to get some help. I would have to talk to some other adults who could help us handle the problem. But I would always talk it over with you first."

DANGEROUS CLIENT BEHAVIOR IN
THE COUNSELING SESSION

Assessing Violence Risk and Responding Appropriately

My client appears angry and agitated, and his referral history reveals that he has been arrested for assault several times. How should I proceed with this client?

Unfortunately, aggressive, acting-out behaviors occur more often within the context of clinical interactions than we would like to think. Statistics indicate that 17% of emergency room patients are violent and that 40% of psychiatrists are physically assaulted at least once during their careers (Tardiff, 1996). Safety should be your first consideration with potentially violent clients. As a counselor intern, keeping safe includes the following:

- Being alert and recognizing potentially threatening clients and/or dangerous situations
- Being aware of the ways your own behavior may either escalate or defuse a client's anger
- Taking precautions to protect yourself and others

There is an increased risk of violence when you are counseling clients who have recently committed a violent act, have been using alcohol heavily, have ingested drugs, are psychotic or have organic brain dysfunctions, have a low IQ and little education, have an unstable residential and employment history, and are males in their late teens or early 20s (Shea, 1998; Truscott & Evans, 2001).

If your client is agitated or is becoming increasingly angry during a counseling session, you should adhere to the following behavioral guidelines:

- Position yourself in front of the client.
- Avoid too much eye contact.
- Do not touch the client.
- Provide extra interpersonal space.
- Speak slowly in a normal conversational tone.
- Carefully explain all your actions in order to avoid arousing suspicion (*e.g.*, "I am going to reach across this table to pick up the pen now.").
- Make sure that the client is able to save face during your discussion.
- Do not challenge the truthfulness of any of the client's statements.
- Encourage the client to verbalize feelings rather than acting them out. (Gabbard, 2000; Shea, 1998).

If you find yourself about to face any new client about whom you have little information, a potentially aggressive client, or a situation that could become dangerous, take steps to ensure your safety. Make certain that help is available and that you will be able to summon assistance quickly and easily. Seat yourself closer to the door; do not allow your client to place himself or herself between you and the door at any time. You may keep the door to the room ajar or open during the session if this seems appropriate. Do not turn your back to the client. That means the client should enter the room ahead of you at the start of the session. At the end of the session, you should exit first, but keep your body turned partially to the client so that you can observe his or her behavior at all times. Trust your intuition. If you feel that you are in danger, explain to your client that you are uncomfortable and need to leave. Notify other staff members immediately, so that everyone can be kept safe and the dangerous client is not left unattended. If a client produces a weapon, leave the room immediately, even if the client is not threatening you personally, and summon help.

Suggested Readings on Dangerous Behavior and Violence

CROWNER, M. L. (2000). *Understanding and treating violent psychiatric patients.* Washington, DC: American Psychiatric Press.

FLANNERY, D. J. (Ed.). *Youth violence: Prevention, intervention, and social policy.* Washington, DC: American Psychiatric Press.

SHEA, S. (1998). *Psychiatric interviewing: The art of understanding.* Philadelphia: Saunders.

SKODOL, A. E. (Ed.). (1998). *Psychopathology and violent crime.* Washington, DC: American Psychiatric Press.

TARDIFF, K. (1996). *Concise guide to assessment and management of violent patients.* Washington, DC: American Psychiatric Press.

OLDER ADULTS

Interacting Therapeutically with Older Adults

My internship site serves some older adults. What can I do in counseling them that would be helpful?

Approximately 15% of the older adults in this country manifest emotional problems, but only 2% receive mental health care (Turner & Helms, 1995). Three factors contribute to the low use of mental health services by older adults. First, there is a widely held, albeit mistaken, belief that depression and other distressing psychological states are a "normal" part of aging. Second, many older adults have limited, fixed incomes and problems with transportation, so that access to mental health care is reduced. Third, many mental health professionals are reluctant to treat older adults, due to misconceptions about aging or personal discomfort with issues of growing old and dying (Butler, Lewis, & Sunderland, 1998; Schwiebert & Myers, 2001).

In counseling older adults, it is helpful to keep in mind that depression and anxiety are *not* inevitable conditions of aging. On the contrary, counseling and psychopharmacological intervention are highly effective in treating mental and emotional disorders of older adults (Binstock, George, Marshall, O'Rand, & Schultz, 2001; Whitbourne, 2001).

Older adults often must cope with loss in many areas. They may be faced with the following:

- Decline in physical strength and vigor, sensory acuity (vision, hearing, and taste), youthful appearance, and sexual potency (in males)

- Loss of power, status, and financial security associated with retirement from career role

- Decreased independence and autonomy due to health problems or limited funds

- Decreased social support system, because friends, siblings, and spouse or significant other may be gone
- Discomfort of certain degenerative illnesses, such as arthritis, associated with aging

In addition, older adults are confronting their own mortality and the fact that "time is running out." They reminisce about the past and try to reconsider experiences and conflicts in order to derive some sense of meaning and continuity in their lives. This process of *life review* is an integral component in the life cycle (Butler, Lewis, & Sunderland, 1998).

During your internship, you can be helpful to your older adult clients by assisting them in their life review. You can listen empathetically to your clients' stories of the past and understand that this narration is not simply "rambling talk" or "resistance to discussing real issues," but rather a valuable and crucial aspect of counseling older adults.

It is imperative to refer your older adult clients to a physician experienced in gerontology, as a prerequisite to counseling. Underlying organic disease causes many cases of depression or impaired thinking. A psychiatric consultation may be required, as well, because psychotropic medications are useful in alleviating many disturbing psychological symptoms (such as insomnia or paranoia) in older adults.

Suggested Readings on Older Adults

CAVANAUGH, J. (1997). *Adult development and aging* (3rd ed.). Pacific Grove, CA: Brooks/Cole.

DUFFY, M. (Ed.). (1999). *Handbook of counseling and psychotherapy with older adults.* New York: John Wiley & Sons.

KENYON, G., CLARK, P., & DE VRIES, B. (Eds.). (2001). *Narrative gerontology: Theory, research, and practice.* New York: Springer Publications.

KNIGHT, B. (1996). *Psychotherapy with older adults* (2nd ed.). Thousand Oaks, CA: Sage.

MYERS, J. E., & SCHWEIBERT, V. L. (1996). *Competencies for gerontological counseling.* Washington, DC: American Psychiatric Press.

TICE, C., & PERKINS, K. (1997). *Mental health issues and aging: Building on the strengths of older people.* Pacific Grove, CA: Brooks/Cole.

WHITBOURNE, S. K. (2001). *Adult development and aging: Biopsychosocial perspectives.* New York: John Wiley & Sons.

GROUP VERSUS INDIVIDUAL THERAPY

Deciding About Group Therapy

One of my clients is shy and has difficulty with interpersonal interactions. How can I decide whether group therapy would be more helpful for her than individual counseling?

Individual therapy does not preclude your client's participation in a group; individual and group therapies are often effective concomitant treatment modalities.

Yalom (1995) emphasizes that groups are suitable for almost all clients, as long as the group's focus, structure, and composition are carefully matched to the client's needs.

Corey and Corey (2002) suggest that group therapy is an especially appropriate format for adolescent clients, because peer interactions are so important during the teenage years. In addition, they write that groups offer an advantage because the adolescent task of separating from adults can sometimes interfere with the development of a positive therapeutic relationship in individual counseling. Group therapy may also be the preferred treatment modality in instances where individual therapy has been frequently disrupted by intensely negative transference (*e.g.*, in some clients with borderline personality disorder), because the presence of other people in the group serves to "dilute" the intensity of the counselor–client relationship somewhat (Yalom, 1995).

Certain general types of clients may not be suitable candidates for group therapy. Usually, clients who have brain injuries, drug or alcohol addictions, paranoid ideation, florid psychotic disorders, or sociopathic behaviors should not be referred to groups (Yalom, 1995). It is also wise to eliminate those clients who are unlikely to attend the group regularly, either because they live too far away geographically from the meeting place, or because they must travel frequently for business (Yalom, 1995). Absences or "dropouts" are disruptive and potentially damaging to the other members of the group.

During your internship, most of your clients will benefit from a combination of group and individual therapies. Facilitating groups is usually an enjoyable learning experience for counselor interns.

Suggested Readings on Group Therapy

COREY, G. (2000). *Theory and practice of group counseling* (5th ed.). Pacific Grove, CA: Brooks/Cole.

COREY, G. (2001). *Theory and practice of counseling and psychotherapy* (6th ed.). Pacific Grove, CA: Brooks/Cole.

COREY, G., & COREY, M. S. (2002). *Groups: Process and practice* (6th ed.). Pacific Grove, CA: Brooks/Cole.

JACOBS, E., MASSON, R., & HARVILL, R. (1998). *Group counseling strategies and skills* (3rd ed.). Pacific Grove, CA: Brooks/Cole.

YALOM, I. (1995). *Theory and practice of group psychotherapy* (2nd ed.). New York: Basic Books.

HOMICIDAL IDEATION

Assessing Risk

One of my clients made a statement about wanting to kill someone. What should I do?

Homicidal ideation always needs to be explored immediately and thoroughly, because there are legal as well as moral implications when clients talk about killing anyone. Even children who express homicidal feelings must be taken very seriously, because there are many cases of children committing murder.

While there is no sure way to predict homicide, the most significant risk factors include past violent behavior, current intent to perform violence, psychosis, being a young male, and heavy use of alcohol and/or drugs (Shea, 1998).

If your client verbalizes homicidal ideation during your internship, you should immediately consult with your site supervisor or another senior clinician on-site. These professionals will decide whether there is a *duty to warn and protect*, meaning that the potential victim(s) should be notified, and whether law enforcement authorities should be involved. As a counselor intern, you do not have enough clinical experience to assess the lethality of homicidal ideation. In case you still feel too uncomfortable to continue working with the client even after the clinician has decided there is no danger of anyone being hurt, discuss the situation and explore alternative treatment options with your supervisor. Make certain to document all interactions and consultations and to let your counseling program supervisor know what is going on.

NONCOMPLIANCE WITH MANDATED COUNSELING

Reporting It

What can I do about a client who is noncompliant with mandated treatment? One of my clients, a 19-year-old male, was required by his college advisor to attend counseling sessions for three months. He came only one time and has not been back.

Some clients, especially those who are mandated to seek counseling by the court or other institutions, will be noncompliant. You should discuss the situation with your supervisor and adhere to agency guidelines. We recommend that you attempt to contact the client, to encourage him or her to return to counseling. However, if you are unsuccessful in reaching the client, or if he or she refuses to return, then you should send a letter to the authorities who initially mandated counseling, explaining the problem and your efforts to resolve it. Your supervisor will need to sign this letter in addition to yourself. Keep a copy of the letter at your internship site and in the client's chart, and keep another copy for yourself. Also, document everything in detail, including dates and times of telephone calls, and keep this documentation in the client's chart.

SERIOUS PSYCHIATRIC DISORDERS

Interacting with Psychotic Clients

I am a counselor intern in an agency that serves clients with serious psychiatric disorders. What should I do if my client seems to be actively psychotic?

If your client population includes individuals with schizophrenia, bipolar illness, or other serious psychiatric disorders, you should discuss with your site supervisor, as soon as you begin your placement, what to do if you encounter a client who is psychotic. Most often, such clients require adjustment of their psychotropic medications. They may have stopped taking their medicine because of uncomfortable side effects. Sometimes, however, you may see a client who comes in with a first episode of psychotic illness. All clients showing signs of active psychotic process should be referred to the psychiatrist for evaluation. They usually will need to be hospitalized, and will be placed on a medication regimen that will require about 10 days to 2 weeks until therapeutic blood levels have been reached.

When interacting with psychotic clients, you should be supportive, calm, and very concrete. Try not to ask too many questions, because many of these clients are suspicious and feel easily threatened and overwhelmed. Do not try to convince clients in an outpatient setting that their hallucinations only exist in their own mind. If you are in an inpatient setting, with staff's approval, you may be able to do gentle reality testing, offering reassurance. For example, you can say: "It must be frightening to hear those voices, but they really are part of your illness and your medication is going to help with that."

Remember that clients who are suffering psychotic illness are not seeing, hearing, interpreting, or understanding the world in the same way that you are. They are out of touch with reality and may have very unpredictable or aggressive behaviors until their psychosis is under control or in remission. Your safety should be your first priority.

Suggested Readings on Serious Psychiatric Disorders

BASCO, M., & RUSH, A. (1996). *Cognitive-behavioral therapy for bipolar disorder.* New York: Guilford.

GABBARD, G. O. (Ed.). (2001). *Treatment of psychiatric disorders* (3rd ed.). Washington, DC: American Psychiatric Publishing.

KAPLAN, H. I., & SADOCK, B. J. (2001). *Kaplan & Sadock's pocket handbook of clinical psychiatry* (3rd ed.). Baltimore: Williams and Wilkins.

KAY, J. (Ed.). (2001). *Integrated treatment of psychiatric disorders.* Washington, DC: American Psychiatric Publishing.

REID, W. H., BALIS, G. U., & SUTTON, B. J. (1997). *The treatment of psychiatric disorders* (3rd ed.) (Revised for DSM-IV). Bristol, PA: Brunner/Mazel.

SHEA, S. (1998). *Psychiatric interviewing: The art of understanding.* Philadelphia: Saunders.

REFERRALS

Finding Resources for Clients
in the Era of Managed Care

How do I refer a client who will need to continue treatment after leaving my agency?

In today's managed care healthcare environment, the third-party payer will usually restrict referrals to clinicians who are on the "provider panel." In these cases, your client will choose a counselor from that list. If the client is not covered by a managed care organization, your agency will outline referral procedures and will usually have a list of community resources available. Your supervisor may be able to offer you suggestions concerning referrals and after-care planning for your client. You can also consult with other clinicians at your site, as well as your counseling program professors. Very often, human service agencies that do not provide the particular service needed by your client will offer names and telephone numbers of other organizations that do. We suggest that you begin assembling your own network of outside therapists and community resources as soon as you can. Collect names of professionals you meet at workshops or meetings; ask colleagues to share names of clinicians they know and respect; and call colleges and universities in your area that offer graduate programs in counseling, social work, or psychology to ask for referral sources.

To begin the referral procedure, explore your client's needs and preferences in detail and try to accommodate your client as much as possible so that he or she will be more likely to follow through on the after-care plans. For example, consider whether your client requires help with emotional or interpersonal issues, family situations, career or work-related problems, or educational concerns. Perhaps your client needs a case manager, who can coordinate services and provide help with such things as transportation or housing, or a psychiatrist, who can prescribe and monitor medications or manage a serious psychiatric disorder. Consider resources such as Alcoholics Anonymous or other community support groups. If your client has a specialized problem related to health or aging, you may call local hospitals, medical schools, or universities to inquire about making referrals to gerontological services, rehabilitation medicine, or psychiatry departments.

The next step in your referral procedure involves choosing several resources that seem to be good possibilities. Share pertinent information with your client, but encourage your client to make the final selection. Explain to your client that he or she may want to try out more than one of these helpers to find the best fit. You may need to offer support for some clients and to actually be there, standing by, as they telephone to set up their first appointments. Finally, complete necessary paperwork and obtain signed forms for release of information, so that you can send client records to the clinician or agency that will be providing care.

CLIENT RESISTANCE

Understanding and Using Resistance

My client has been arriving late for our sessions or sometimes missing her appointments. When she does come for counseling, she discusses everything except her own issues. She often will talk about her neighbors, the weather, or a television program in great detail. I know this client is being resistant, but I am not sure why, or what to do about it.

Resistance refers to "any idea, attitude, feeling, or action (conscious or unconscious) that fosters the status quo and gets in the way of change" (Corey, 2001,

p. 94). Resistance comes into play when the client consciously or unconsciously tries to avoid painful, threatening, or unpleasant thoughts and feelings. You can therefore understand resistance as a defense against anxiety and as a protective coping mechanism during counseling. As a novice counselor, you will need to remember that your client is not simply being "uncooperative" and that the resistant behavior is not meant to frustrate you (Corey, 2001; Kahn, 1997). Resistance during counseling should be appreciated as a means to understand your client's inner world more deeply. Your client's resistance in counseling provides a clue to the way he or she approaches problems in daily life (Corey, 2001).

One strategy in addressing your client's resistance is encouraging your client to explore the source of the resistant behaviors. When your client becomes aware of the meaning of and need for such behaviors as being late, skipping sessions, remaining silent for long periods of time, asking questions about your personal life, and making "small talk," he or she will be more ready to acknowledge and master the underlying painful issues that he or she has been avoiding. Often, just your empathic comment may help your client overcome resistance to discussing difficult issues: "It's hard for you to get started talking about your father, isn't it?" Working through resistance allows the client the freedom to address important issues and make positive life changes.

Suggested Readings on Client Resistance

COREY, G. (2001). *Theory and practice of counseling and psychotherapy* (6th ed.). Pacific Grove, CA: Brooks/Cole.

GABBARD, G. (2000). *Psychodynamic psychiatry in clinical practice* (2nd ed.). Washington, DC: American Psychiatric Press.

KAHN, M. (1997). *Between therapist and client: The new relationship* (2nd ed.). New York: Freeman.

SHEA, S. (1998). *Psychiatric interviewing: The art of understanding* (2nd ed.). Philadelphia: Saunders.

URSANO, R., SONNENBERG, S., & LAZAR, S. (1998). *Concise guide to psychodynamic psychotherapy: Principles and techniques in the era of managed care.* Washington, DC: American Psychiatric Press.

SUICIDAL BEHAVIOR

Assessing Risk

My client appears to be very depressed and has made the comment, "I really think I'd be better off dead." How do I decide what to do next?

All suicidal statements need to be taken seriously and thoroughly evaluated. About two-thirds of those people who express suicidal feelings eventually do kill themselves, and completed suicide occurs in more than 15% of persons suffering from depression (Gabbard, 2001). There are more than 25,000 suicides in the United States every year (Shea, 1999). Older men, adolescents, and persons suffering from severe major depression or psychosis with religious preoccupation are at the most lethal risk of killing themselves (Shea, 1998). However, suicide occurs across the life span, in both males and females, in all socio-economic environments (Shea, 1998, 1999).

Some interns are afraid that if they ask a question such as, "Have you been thinking about hurting or killing yourself?" they will be giving the client the idea. *Asking about suicidal thoughts will not encourage your client to commit suicide.* Most of the time, clients are relieved to be asked, because they have ambivalent feelings and thoughts about suicide, and want to be helped to feel better. As a counselor intern, you should attempt to elicit suicidal ideation with every client, and you should certainly explore and assess all suicidal thoughts or statements to determine lethality, so that you can take action to help ensure your client's safety.

The client who verbalizes suicidal wishes is expressing the intensity of his or her feelings of hopelessness, helplessness, loneliness, or other pain. You should respond with empathy and sensitivity to your client's distress, while evaluating the following hierarchy of risk factors:

- Intent
- Plan
- Means

First, ask about the client's intent to kill himself or herself: "Have you been feeling like hurting or killing yourself?" It is also important to inquire about the intensity of suicidal thoughts or feelings: "How often do you think about killing yourself?" or "How long did you stand on the bridge wondering whether to jump?"

Second, ask whether the client has a concrete plan: "Have you been thinking of how you would kill yourself? Tell me about what you would do."

Third, ask whether the client has the means to carry out this plan: "Do you have access to a gun?" or "How many of your pills have you saved up by now?" Clients who have suicidal ideation along with intent to kill themselves, a definite plan, and the means to execute this plan present a very high risk.

We would also like to note several other situations that should alert you to suicidal risk in your client:

- The client has been taking antidepressants for several weeks, to relieve symptoms of a mood disorder, and has started to feel better. Often, paradoxically, when clients begin to feel better, they have enough energy to carry out their plans to commit suicide.
- The client has made previous suicidal gestures or attempts.
- The client has no social support system, or *believes* that he or she has no social support system. Sometimes the client's perceptions regarding the availability of others are incorrect, but keep in mind that it is this belief that is acted upon.
- Family members, friends, classmates, or coworkers have committed suicide, either in the past or more recently. Anniversary dates of those events may be especially difficult.

- The depressed client appears suddenly calm or speaks of feeling relieved, of having found a "solution" to his or her pain. Always ask about the source of relief.

- The client gives away possessions, or makes arrangements concerning legal or financial affairs.

- The client is an adolescent who is very distraught. Many adolescents are impulsive and have not yet developed either good judgment or the ability to foresee the long-term consequences of acting on their immediate feelings.

- The client has been suffering from a serious medical illness or chronic health problem.

- The client has a diagnosis of schizophrenia, major depression, alcohol or drug dependency, or borderline personality disorder.

- Your intuition tells you that something is not right, or that the client is upset enough to hurt himself or herself.

Once you have identified, explored, and assessed suicidal ideation and have determined that there is significant risk, what should you do next? First, be certain to let your supervisor know about the situation before you allow the client to leave the office. Depending on the circumstances and your experience, your supervisor may want to interview the client or to review your plans. Next, mobilize the client's own resources, such as religious values, spiritual beliefs, family, friends, and the healthy part of the self that is seeking help. Ask your client what has been helpful in resisting the impulse to kill himself or herself so far, and emphasize the importance of these factors. Explain that depression and suicidal feelings are time-limited and treatable, and that the client will be able to get help to feel better.

If you and your supervisor ascertain that it is safe for the client to go home, secure a signed behavioral contract in which the client agrees he or she is able to stay safe, and will telephone for help or go to the hospital emergency room if suicidal feelings become unmanageable. Provide a 24-hour telephone number for your client, in accordance with agency regulations, such as a crisis hot-line, psychiatric emergency service, or clinician on call. Maintain a close, supportive relationship with the client and increase the frequency of your counseling sessions, even seeing the client daily if necessary, until the crisis subsides.

If the client cannot agree to keep himself or herself safe, or if your gut feeling tells you that the client is still in trouble, use other resources. Ask the client if there are family members or friends who can be called to help. Consult with your supervisor or other clinicians on-site. If hospitalization is indicated, follow agency procedures. Try to encourage voluntary hospitalization for the seriously suicidal client by discussing the need for "getting all the help we can to keep you safe." Unfortunately, involuntary hospitalization is sometimes necessary. If this is the case, your site supervisor should be close by to assist. Follow agency guidelines, hospital policies, and county or state regulations carefully. In all instances, carefully document the client's mood, affect, cognition, behavior, and exact statements, as well as your interventions and plans.

Suggested Readings on Suicidal Behavior

BERMAN, A., & JOBES, D. (1996). *Adolescent suicide: Assessment and intervention.* Washington, DC: American Psychological Association.

ELLISON, J. M. (Ed.). (2001). *Treatment of suicidal patients in managed care.* Washington, DC: American Psychiatric Press.

JACOBS, D. G. (Ed.). (1999). *The Harvard Medical School guide to suicide assessment*

and intervention. San Francisco: Jossey-Bass.

SHEA, S. (1998). *Psychiatric interviewing: The art of understanding* (2nd ed.). Philadelphia: Saunders.

SHEA, S. (1999). *The practical art of suicide assessment: A guide for mental health practitioners and substance abuse counselors.* New York: John Wiley & Sons.

TRANSFERENCE AND COUNTERTRANSFERENCE

Understanding and Using Transference and Countertransference

I have been counseling a client for several weeks and I thought that we had a good therapeutic relationship. Recently, however, she has become angry with me during our sessions and accuses me of rejecting her. I actually like this client very much, but now I often want to avoid our sessions because I find them unpleasant, and I have begun distancing myself emotionally from my client. I am not sure what is going on or how to repair our relationship.

You are experiencing psychological phenomena known as *transference* and *countertransference*. These concepts constitute central dynamics in psychoanalytic therapy; however, an appreciation of the importance of transference and countertransference is pertinent to your understanding of all interpersonal interactions, including all therapeutic relationships.

Transference refers to the transfer of feelings, attitudes, fears, wishes, desires, and perceptions that belonged to our past relationships onto our current relationships, so that the people in our present lives become the focus of these thoughts and emotions from long ago (Corey, 2001). Transference also refers to the tendency of the client to transfer feelings from other current relationships onto the counselor, thereby reexperiencing these feelings in the counseling session.

Transference occurs for several reasons. First, we learn to view our environment and other people's behavior in light of the experiences and relationships we have had while growing up (Corey, 2001). Second, the human mind constantly compares incoming information with both conscious and unconscious memories in an effort to find some pattern match and to organize new information in a meaningful way (Eisengart & Faiver, 1996). Third, there seems to be a psychological need in all people to repeat and try to master experiences and relationships that were difficult or emotionally painful in the past (Ursano, Sonnenberg, & Lazar, 1998). Therefore, our perceptions of the present are frequently colored by our memories of past situations, and transference occurs commonly in many human experiences.

Because counseling often evokes powerful feelings through discussion of conflicts from both the past and the present, transference in the context of a counseling relationship can become especially intense and can feel very real to the client. In the question posed above, your client is experiencing your behavior as rejecting, just as she experienced rejection from some other significant person in her life, perhaps from a parent. Your client also may be reexperiencing the same feelings of pain and anger toward you as she did toward that other person, long ago. You can help her by demonstrating that, in reality, you do care very much for her, that you do want to "be there" for her, and that you are trying hard to understand her feelings. You may also explain the concept of transference and encourage her to explore where her feelings may be originating.

Countertransference is similar to transference and refers to the counselor's emotional responses to the client. Countertransference may be due to the unconscious feelings, thoughts, and conflicts from the counselor's own past that are evoked by the present clinical situation. However, countertransference can also be interpreted as a reality-based response to the client's transference, and it can be used by the counselor to gain insight.

Ursano et al. (1998) explain that countertransference can be experienced in two ways. In the first, the counselor experiences the same emotions as the client and feels deep empathy for the client. In the second, the counselor experiences feelings toward the client that are similar to feelings experienced by another important individual in the client's life, past or present. You seem to be experiencing the second type of countertransference reaction, because you find yourself wanting to avoid sessions and you are emotionally distancing yourself from your client. Thus, you are likely experiencing the same rejecting feelings toward your client as another person in her life displayed toward her long ago. Understanding where your feelings are coming from may help you to get in touch with your warm and accepting feelings for the client, so that you can reestablish your previous good relationship.

Suggested Readings on Transference and Countertransference

COREY, G. (2001). *Theory and practice of counseling and psychotherapy* (6th ed.). Pacific Grove, CA: Brooks/Cole.

GABBARD, G. O. (Ed.). (1999). *Countertransference issues in psychiatric treatment.* Washington, DC: American Psychiatric Press.

GABBARD, G. (2000). *Psychodynamic psychiatry in clinical practice* (2nd ed.). Washington, DC: American Psychiatric Press.

KAHN, M. (1997). *Between therapist and client: The new relationship* (2nd ed.). New York: Freeman.

ST. CLAIR, M. (1997). *Object relations and self-psychology: An introduction* (2nd ed.). Pacific Grove, CA: Brooks/Cole.

URSANO, R., SONNENBERG, S., & LAZAR, S. (1998). *Concise guide to psychodynamic psychotherapy: Principles and techniques in the era of managed care.* Washington, DC: American Psychiatric Press.

CONCLUDING REMARKS

In this chapter, we have attempted to provide guidelines for handling some of the professional dilemmas and problematic counseling situations you may face during your internship. We encourage you to consult with both your site supervisor and your program supervisor and to use other available resources whenever you need help in resolving a challenging problem during your internship.

REFERENCES

BARKLEY, R. A. (1998). *Attention deficit hyperactivity disorder: A handbook for diagnosis and treatment.* New York: Guilford Press.

BINSTOCK, R. H., GEORGE, L. K., MARSHALL, V. W., O'RAND, A. M., & SCHULTZ, J. H. (Eds.). (2001). *Handbook of aging and the social sciences.* San Diego, CA: Academic Press.

BREMS, C. (2000). *Dealing with challenges in psychotherapy and counseling.* Pacific Grove, CA: Brooks/Cole.

BREMS, C. (2001). Counseling clients with underlying medical problems. In E. R. Welfel & R. E. Ingersoll (Eds.), *The mental health desk reference* (pp. 10–18). New York: John Wiley & Sons.

BUTLER, R. N., LEWIS, M. I., & SUNDERLAND, T. (1998). *Aging and mental health: Positive psychosocial and medical approaches.* Boston: Allyn & Bacon.

COREY, G. (2001). *Theory and practice of counseling and psychotherapy* (6th ed.). Pacific Grove, CA: Brooks/Cole.

COREY, G., & COREY, M. S. (1997). *Groups: Process and practice* (5th ed.). Pacific Grove, CA: Brooks/Cole.

COREY, G., & COREY, M.S. (2002). *Groups: Process and practice* (6th ed.). Pacific Grove, CA: Brooks/Cole.

COREY, G., COREY, M. S., & CALLAHAN, P. (1998). *Issues and ethics in the helping professions* (5th ed.). Pacific Grove, CA: Brooks/Cole.

DACEY, J., & TRAVERS, J. (1999). *Human development across the lifespan* (4th ed.). Boston: McGraw-Hill.

DANZINGER, P. R. (2001). Defining and recognizing elder abuse. In E. R. Welfel & R. E. Ingersoll (Eds.), *The mental health desk reference* (pp. 478–483). New York: John Wiley & Sons.

DUFFY, M. (Ed.). (1999). *Handbook of counseling and psychotherapy with older adults.* New York: John Wiley & Sons.

DULCAN, M., & MARTINI, C. (1998). *Concise guide to child and adolescent psychiatry* (2nd ed.). Washington, DC: American Psychiatric Press.

EISENGART, S., & FAIVER, C. (1996). Intuition in mental health counseling. *Journal of Mental Health Counseling, 18*(1), 41–52.

ERIKSON, E. (1963). *Childhood and society* (2nd ed.). New York: Norton.

GABBARD, G. (2000). *Psychodynamic psychiatry in clinical practice* (3rd ed.). Washington, DC: American Psychiatric Press.

GABBARD, G. O. (Ed.). (2001). *Treatment of psychiatric disorders* (3rd ed.). Washington, DC: American Psychiatric Publishing.

HANNA, F. J., HANNA, C. A., & KEYS, S. G. (1999). Fifty strategies for counseling defiant, aggressive adolescents: Reaching, accepting, and relating. *Journal of Counseling and Development, 77,* 395–404.

KAHN, M. (1997). *Between therapist and client: The new relationship* (2nd ed.). New York: Freeman.

KALICHMAN, S. (1999). *Mandated reporting of suspected child abuse: Ethics, law, and policy* (2nd ed.). Washington, DC: American Psychological Association.

KALICHMAN, S. (2001). Reporting suspected child abuse. In E. R. Welfel & R. E. Ingersoll (Eds.), *The mental health desk reference* (pp. 471–477). New York: John Wiley & Sons.

KEARNEY, C. A. & LINNING, L. M. (2001). Treating anxiety disorders in children. In E. R. Welfel & R. E. Ingersoll (Eds.), *The mental health desk reference* (pp. 177–182). New York: John Wiley & Sons.

McWHIRTER, B. T., & BURROW, J. J. (2001). Assessment and treatment recommendations for children and adolescents with depression. In E. R. Welfel & R. E. Ingersoll (Eds.), *The mental health desk reference* (pp. 199–204). New York: John Wiley & Sons.

MORRISON, J. R. (1997). *When psychological problems mask medical disorders.* New York: Guilford.

REINECKE, M. A., RYAN, N. E., & DUBOIS, D. L. (1998). Cognitive-behavioral therapy of depression and depressive symptoms during adolescence. *Journal of the American Academy of Child and Adolescent Psychiatry, 37,* 214–222.

SCHWIEBERT, V. L., & MYERS, J. E. (2001). Counseling older adults. In E. R. Welfel & R. E. Ingersoll (Eds.), *The mental health desk reference* (pp. 320–325). New York: John Wiley & Sons.

SHEA, S. (1998). *Psychiatric interviewing: The art of understanding* (2nd ed.). Philadelphia: Saunders.

SHEA, S. (1999). *The practical art of suicide assessment: A guide for mental health practitioners and substance abuse counselors.* New York: John Wiley & Sons.

TARDIFF, K. (1996). *Concise guide to assessment and management of violent patients.* Washington, DC: American Psychiatric Press.

THOMPSON, C., & RUDOLPH, L. (2000). *Counseling children* (5th ed.). Pacific Grove, CA: Brooks/Cole.

TRUSCOTT, D., & EVANS, J. (2001). Responding to dangerous clients. In E. R. Welfel & R. E. Ingersoll (Eds.), *The mental health desk reference* (pp. 271–278). New York: John Wiley & Sons.

TURNER, J. S. & HELMS, D. B. (1995). *Lifespan development* (5th ed.). Chicago: Holt, Rinehart and Winston.

URSANO, R., SONNENBERG, S., & LAZAR, S. (1998). *Concise guide to psychodynamic psychotherapy: Principles and techniques in the era of managed care.* Washington, DC: American Psychiatric Press.

WHITBOURNE, S. K. (2001). *Adult development and aging: Biopsychosocial perspectives.* New York: John Wiley & Sons.

WICKRAMASEKERA, I., DAVIES, T. E., & DAVIES, S. M. (1996). Applied psychophysiology: A bridge between the biomedical model and the biopsychosocial model in family medicine. *Professional Psychology: Research & Practice, 27*(3), 221–233.

YALOM, I. (1995). *Theory and practice of group psychotherapy* (4th ed.). New York: Basic Books.

10

Ethical Practice
in Counseling

The term *ethical* means that a behavior or practice conforms to the specialized rules and standards for proper conduct of a particular profession. Ethics, therefore, is "a fundamental and defining aspect of professionalism" (Gibson & Pope, 1993, p. 330). The code of ethics is basic to the identity of any professional organization because it provides behavioral norms and guidelines that structure each member's professional activities as well as ensure uniformly high professional standards among all members.

Counselors-in-training have an obligation to know and to adhere to the American Counseling Association (ACA) *Code of Ethics and Standards of Practice* (1995) in addition to any pertinent state codes and laws. These standards spell out distinct behavioral parameters within which all counselors, including yourself as a counselor intern, must operate. The ACA *Code of Ethics* is divided into eight sections, each describing one aspect of proper professional behavior.

Counselor interns also need to be aware of and to function within the boundaries of their agency or institutional policies, provided that these policies do not conflict with the ACA *Code of Ethics*. If you perceive a possible conflict, realize that the ACA *Code of Ethics* takes precedence over agency or institutional policies. We suggest that you discuss the issue with your agency supervisor and university supervisor to try to resolve the problem. If the issue is not resolved at this level, then you may wish to contact the ACA or the ethics committee of your state licensure or certification board or your state division of the ACA for further advice or assistance. Finally, as an intern and future professional counselor, you should always conduct your counseling activities in accordance with state and local laws and regulations. These rules can provide you with important professional backup and a sense of security in dealing with uncomfortable or equivocal professional situations.

ETHICAL DILEMMAS

Ethical codes provide guidelines and principles for ethical conduct. Realistically, though, as counselors we will encounter many professional situations that are not specifically addressed by the ACA *Code of Ethics* (Corey, 2001; Corey, Corey, & Callanan, 1998; Gibson & Pope, 1993). Furthermore, certain clinical circumstances may involve conflicting interests or multiple ethical principles that leave us with exceedingly difficult choices. "It is the translation of a code's principles into practical directions for conduct that is the greatest challenge for most of us" (Gibson & Pope, 1993, p. 330). Herlihy and Corey (1996) suggest these five general ethical principles of conduct for counselors. They include:

1. *Autonomy,* which refers to assisting clients in developing a sense of empowerment, independence, self-determination, and an acceptance of different values

2. *Nonmaleficence,* meaning to do no harm, even inadvertently

3. *Beneficence,* by promoting mental health and wellness

4. *Justice,* or a commitment to fairness, quality of service, allocation of time and resources, establishment of fees, and access to services

5. *Fidelity,* which implies that counselors honor commitments to clients, students, and supervisees

Professional dilemmas frequently occur when personal morality conflicts with professional ethics. Our client may reveal to us that he or she is engaging in some behavior that we personally believe to be wrong or even immoral, yet we are bound by our ethical guidelines and the rules of professional confidentiality and cannot take any action outside of the treatment situation unless harm is threatened to self or others. The result can be uncomfortable cognitive dissonance.

One counselor intern agonized over a client who reported driving drunk several nights a week. Although he did encourage his client to work on this issue during their sessions, the intern told us, "If I could just be a private citizen and not a professional counselor, I would make a telephone call to report this person immediately! I feel a moral responsibility to keep her off the road." Another area of potential professional quandary involves choice of "client." For example, if you feel that a client is antisocial, presents a possibility of harm to others, or poses an outright danger to society, is that person your client or is the system or the general public your client? Where does your responsibility lie? One counselor intern described her dilemma when a client who was HIV positive disclosed that his partners were unaware of this test result and that he frequently did not adhere to safe sex practices. This intern wondered about her duty to warn and protect the partners versus her duty to protect client confidentiality. Another counselor intern described a situation in which she was counseling an undergraduate student with a history of violent, aggressive behavior, which was unknown to university administrators. This undergraduate was granted permission to live in a dormitory on campus. The counselor intern worried about the safety of the other students yet felt that she could not warn university officials because the client had made no specific threats to hurt anyone.

GETTING HELP IN RESOLVING ETHICAL PROBLEMS

During your internship, and throughout your counseling career, you will grapple with difficult decisions as you try to apply the ACA *Code of Ethics* to ambiguous or complex counseling situations. We recommend using a variety of resources to help resolve your concerns, including supervisors and colleagues, the ACA Ethics Committee, state licensure or certification boards, and professional journals and texts that address ethics in counseling. In their book *Issues and Ethics in the Helping Professions,* Corey, Corey, and Callanan (1998) analyze many ethical dilemmas counseling professionals may confront. We encourage you to read this text thoroughly to help prevent violation of accepted practice standards and to refer to it in any questionable professional circumstances, as well.

As a counseling professional, you will find that your own values and morals, as well as your personal conceptualization of ethical counseling behavior, influence your interpretation and application of the ACA *Code of Ethics.* Scott (1998) suggests six approaches to making ethical choices:

1. Consider moral principles in determining what is right and wrong.
2. Implement pragmatic moral strategies based upon the anticipated outcome.
3. Assess the situation in the context of accepted and appropriate behavior.
4. Use your intuition regarding what is right or wrong.
5. Strive to gain the appropriate benefits for the client.
6. Look for the greater good; consider what is best for the long term.

Keeping these factors in mind may clarify your understanding of those ethical dilemmas in counseling where the ACA *Code of Ethics* does not seem to provide definitive answers.

A BRIEF SUMMARY OF THE
ACA *CODE OF ETHICS*

The first step in upholding the ethical standards and practices of the counseling profession during your internship is gaining a working knowledge of the ACA *Code of Ethics.* Many counseling students tend to avoid thoroughly reading or examining the ACA *Code of Ethics.* They feel that it is too complicated, too lengthy, and inaccessible because it is often relegated to an appendix at the back of a book rather than being included within the main text. In addition, the ACA *Code of Ethics* may seem somewhat remote from their immediate needs and frame of reference as graduate students rather than as counseling practitioners. However, as a counselor intern, your role has evolved from that of primarily a student to that of essentially a clinician. We have therefore provided the following summary of the ACA *Code of Ethics* in an effort to make it more "user-friendly." It is interesting to note that the code is grounded in the eight CACREP (Council for Accreditation of Counseling and Related Educational Programs) core areas of study. We strongly recommend that you set aside the time, early in your internship, to study the ACA *Code of Ethics* in its original, unabbreviated form and keep a copy handy for reference. A solid understanding of the ACA *Code of Ethics* will provide you with a knowledge base as you struggle to apply ethical principles to the diversity of challenging situations you will encounter during your counseling internship and your counseling career.

Section A: The Counseling Relationship

Counselors always respect the rights, the freedom, the integrity, and the welfare of the client, whether in an individual relationship or in a group setting. Counselors believe that clients have the potential to grow and change and strive to empower them in this endeavor. Counselors develop with clients measurable, observable, and realistic treatment plans that provide considerable flexibility in the therapeutic process.

Counselors value the importance of the family as both a support system and a resource for clients, while always protecting the individual privacy rights of their clients. Counselors need to understand the strengths and limitations of their clients when career and employment needs become factors in the treatment process. Clients' multicultural diversity must always be respected and understood; at the same time, counselors must gain insight into their own unique identities in relating to clients.

Counselors explain the goals, structure, expectations, rules, treatment techniques, proposed treatment outcomes, and any limitations of counseling, including professional competencies, to clients at the outset of the counseling relationship. Clients (or guardians or parents of minors) can terminate the counseling relationship if they perceive no interest and benefit. Counselors do not enter a professional relationship with a client who is already involved in another counseling relationship unless proper notification is given and a request to terminate and transfer is initiated.

Counselors maintain professional boundaries and do not make their personal needs an issue. Dual relationships (such as both counselor and teacher, supervisor and supervisee, or counselor and friend) should be avoided, and referral should be made. Counselors view sexual intimacies or relationships with clients and former clients as inappropriate and unethical. Boundaries must be clear and roles defined so as to avoid any conflict.

Counselors assume responsibility for protecting group members from psychological or physical suffering resulting from participation in the group. Counselors protect group members by screening prospective members, by maintaining an awareness of group interactions and relationships, and by making available professional assistance for members who require such intervention, both during and after group sessions. Group members need to be apprised of the goals and objectives, rules, roles, and projected treatment outcomes. Counselors advise group members about respecting rules of confidentiality and avoid the dual roles of performing both individual and group counseling.

Counselors assist clients with necessary financial arrangements related to the counseling relationship, realizing that fees are an integral element of the therapeutic process. When counselors have to interrupt their service delivery to clients, they must ensure the continuity of treatment. In addition, counselors advise clients when they are unable to be of professional help and then assist these clients in finding alternatives.

When using computers as part of counseling services, counselors take precautions to make certain that the client is intellectually, physically, and emotionally capable of using a computer, that the computer program is appropriate and understandable, and that follow-up counseling is provided. Counselors who design computer software ensure that self-help/stand-alone software has actually been designed as such. Counselors designing such software also provide descriptions of expected outcomes, suggestions for use, inappropriate applications, possible benefits, and a manual defining any other essential data. Computer programs, methods, and techniques adhere to applicable professional standards, ethics, and guidelines.

Section B: Confidentiality

Counselors maintain the standard of confidentiality as they work to establish the therapeutic relationship. This standard is violated only when the safety and well-being of the clients, those individuals around them, and the general community are in jeopardy. Counselors inform clients of the limitations of confidentiality and identify situations when confidentiality may be breached, including court orders, treatment team approaches, clinical and fiscal audits, and quality assurance reviews.

Confidentiality is preserved when working with groups, families, and minor or incompetent clients. Client records meet the regulatory and best-practice standards in order to safeguard the integrity of the individual. Clients have the right to access their records with the proviso that they are competent in understanding their records or have an advocate to interpret the content.

Counselors who use client data for research purposes ensure client anonymity. Counselors who are consultants share client information only with individuals who are clearly concerned with and responsible for the case. Individuals must sign releases of information when records or other client data are shared in any form.

Section C: Professional Responsibility

Counselors operate within their scope of competencies, education and training, and professional experience. Counselors are committed to continuing education and awareness in working with diverse populations, diagnoses, and theories and techniques of treatment. Counselors accept employment opportunities that are congruent with their education and training and continually solicit feedback about the quality of their professional work. Counselors seek the guidance and input of other colleagues when questioning ethical obligations or concerns of professional practice. Counselors avoid offering or accepting counseling responsibilities when impaired either physically or emotionally. Counselors accurately present their credentials and scope of practice to the general public and their clients. They rely on the rules and guidelines of their respective licensure or certification boards. Counselors immediately correct any omissions in education, training, licensure, or certification.

Counselors provide their services to individuals regardless of age, color, culture, disability, ethnic group, gender, race, religion, sexual orientation, or socioeconomic status. Counselors respect differences in colleagues' therapeutic approaches. They endeavor to take into account the varying traditions and practices of other professional groups with whom they work.

Section D: Relationships with Other Professionals

Counselors define and describe their professional roles and responsibilities with employers, employees, colleagues, and supervisors. These delineations are reflected in working agreements, clinical relationships, and issues of confidentiality; they also adhere to professional standards. Peer review of the counselor's quality of work, as reflected in ongoing professional evaluations, provides professional feedback that improves service delivery. Continuing in-service education is central to the refinement of clinical knowledge and skills.

Counselors' personal and professional practices demonstrate respect for the rights and well-being of colleagues and at the same time maintain the highest level of professional service. Counselors in administrative or supervisory capacities ensure that staff members operate within their level of competency. Counselors maintain the highest standard of professional conduct in their relationships with both clients and affiliated agencies. Counselors abide by the policies and procedures of their agencies. In addition, counselors understand the need for ethical reassessment and modification of policies and procedures in order to promote better service.

Section E: Evaluation, Assessment, and Interpretation

Counselors use educational and psychological assessment techniques in the best interests of client welfare. Counselors do not misuse test results and interpretations and endeavor to prevent others from doing so. Counselors strive to respect clients' rights to know test results, interpretations, and the basis for conclusions and recommendations. Counselors are aware of limits of competency, performing only those testing and assessment services in which they have been adequately trained.

Counselors consider reliability and validity data; the currency or obsolescence of tests; proper administration of any technique utilized; and awareness of socioeconomic, cultural, and ethnic factors. Plake and Impara's (2001) *The Fourteenth Mental Measurements Yearbook* and Murphy, Impara, and Plake's (1999) *Tests in Print V* are excellent resources for understanding standardized tests.

Counselors are responsible for ensuring the appropriate application, scoring, interpretation, and use of assessment instruments. Before administering an assessment instrument, counselors clearly explain the nature and purpose of the assessment as well as how the results will be used. Counselors are careful in providing the proper diagnoses. Counselors are careful in conducting clinical interviews and selecting assessment techniques. Counselors strive to maintain the integrity and security of tests and other assessment instruments in compliance with legal and contractual requirements.

Section F: Teaching, Training, and Supervision

Counselors responsible for training counselor education students and counselor supervisees are competent practitioners and teachers. They model the highest ethical standards of the profession, maintaining appropriate professional and social boundaries. Counselor educators are adequately trained in supervision methods. Further, they ensure that students and supervisees assume a professional role in providing services.

Counselor educators advise prospective students concerning program requirements, including the scope of professional skills development. They provide a comprehensive view of the counseling profession, including employment opportunities, before admission to the counselor education training program. Counselor educators establish training sites that integrate academic study and supervised practice. These sites expose trainees to diverse theoretical orientations and populations. This exposure affords the trainees the opportunity to increase their competencies and knowledge base.

Students and supervisees are provided clear guidelines pertaining to competency levels expected and subsequent appraisal and evaluation methods for both didactic and experiential training components. Periodic performance appraisal with feedback is provided. Counselor educators communicate expectations of trainees' roles and responsibilities to their trainees. Care is taken in confirming that site supervisors are qualified to provide supervision and will adhere to the ethical standards of this role. Counselor educators must avoid duplicity of roles as field supervisor and training program supervisor. Counselor educators do not counsel their students or supervisees; in such cases, referral is made to another practitioner.

Section G: Research and Publication

Counselors performing research with human subjects adhere to ethical principles and regulations consistent with federal and state laws and scientific standards. At all times counselors protect the rights of research participants. Counselors avoid anything that could injure their subjects. The principal researcher holds the primary responsibility for ensuring the maintenance of ethical practice. However, all counselors involved in a study are aware of ethical principles and assume responsibility for their actions. Counselors inform subjects of the purpose, procedures, risks, and benefits of the investigation if possible. Researchers carefully disguise subjects' identities. Participant data are kept in strictest confidence. Further, counselors honor all commitments to subjects.

Counselors report research findings in an ethical manner, noting any conditions that may misleadingly affect results. Original research data are reported in order to allow for replication. Previous research findings are appropriately and objectively cited. Counselors publishing material are careful to credit the work of others, such as through the use of citations, references, footnotes, acknowledgments, or joint authorship. Counselors jointly performing research are responsible for the accuracy and thoroughness of the information they contribute. In addition, counselors who collaborate incur an ethical responsibility to other collaborators to honor agreed-upon commitments. Counselors do not submit work to more than one journal or publisher at a time. Counselors obtain publisher permission for reprinting manuscripts.

Section H: Resolving Ethical Issues

Counselors are familiar with the ACA *Code of Ethics,* realizing that ignorance or misunderstanding of an ethical responsibility is not an acceptable defense against a charge of unethical conduct. When counselors have specific ethical questions, they confer with other counselors or appropriate authorities. When ethical issues emerge regarding their affiliated organization, counselors specify the nature of such conflict, initially attempting to resolve the issue with their supervisor or other responsible official. Counselors may hypothetically discuss ethical issues and concerns with colleagues. Counselors take cautions so as not to initiate, participate in, or encourage the filing of unwarranted or harmful ethical complaints. Counselors cooperate in any proceedings to investigate ethical complaints at all levels.

A MODEL FOR ETHICAL COUNSELING PRACTICE

Ethical behavior requires more than a familiarity with the profession's code of ethics. Counselors also need to develop a personal ethical sense that involves reflection and insight in assuring the best possible service delivery to their clients, students, or supervisees (Herlihy & Corey, 1996). The willingness to accept possible consequences of ethical actions may be considered a moral decision.

Welfel (2002) presents a process model for ethical decision-making composed of 10 stages. They include:

1. *Sensitivity* to the moral dimensions of counseling, which encompass not only professional ethics, but also personal principles and philosophy consistent with the profession.
2. *Identification* of the type or category of dilemma and alternative courses of action.
3. *Referral* to the *Code of Ethics* and professional guidelines for guidance.
4. *Examination* of the relevant federal and state regulations and case law for additional guidance.
5. *Examination* of relevant ethics literature for perspective.
6. *Application* of fundamental philosophical principles and theories to the situation.
7. *Consultation* with colleagues about the dilemma. Consultation is always done within the framework of the ethical standards as well.
8. *Deliberation*. Welfel suggests that this be personal deliberation, through which the counselor considers alternatives and develops a plan for action.
9. *Informing* appropriate persons, such as supervisors, and implementing the decision.
10. *Reflection* on the action, which provides counselors an assessment and affirmation of their ethical decision-making process.

A WORD ON THE ETHICS OF INTERNET COUNSELING

Counseling services are increasingly offered via the Internet. This modality presents unique ethical concerns such as the elimination of state and national boundaries, lack of counseling regulatory agencies' involvement, and the usual lack of visual input (Hackney, 2000). Many counselors agree that the Internet may be beneficial, yet recognize the potential for problems (Sampson, Kolodinsky, & Greeno, 1997). We encourage counselors to review the *Ethical*

Standards for Internet Online Counseling, which were approved by the ACA Governing Council, in October 1999. Counselor trainees can review the ethics code for online counseling at www.counseling.org.

CONCLUDING REMARKS

You will surely confront a broad scope of ethical issues and dilemmas during your counseling internship and throughout your professional career. A knowledge of the ACA *Code of Ethics* gives you a foundation for your ethical decisions; however, you will need to examine your own values closely as you search for acceptable solutions to some ethical problems. Herlihy and Corey (1996) view the *Code of Ethics* as a necessary and vital core component of the education and training of counselors, resulting in sound professional conduct, accountability, and improved practice. In addition, they feel ethical codes provide a combination of rules and utilitarian principles in ways that require interpretation. We encourage you to discuss ethical questions and issues with your colleagues, supervisors, and professors and to refer to the professional journals and books that address counseling ethics as you search for those answers.

REFERENCES

COREY, G. (2001). *Theory and practice of counseling and psychotherapy* (6th ed.). Belmont, CA: Brooks/Cole.

COREY, G., COREY, M., & CALLANAN, P. (1998). *Issues and ethics in the helping professions* (5th ed.). Pacific Grove, CA: Brooks/Cole.

GIBSON, W., & POPE, K. (1993). The ethics of counseling: A national survey of certified counselors. *Journal of Counseling and Development, 71,* 330–336.

HACKNEY, H. (2000). *Practice issues for the beginning counselor.* Needham Height, MA: Allyn & Bacon.

HERLIHY, B., & COREY, G. (1996). *ACA ethical standards casebook* (5th ed.). Alexandria, VA: American Counseling Association.

MURPHY, L. L., IMPARA, J. C., & PLAKE, B. S. (Eds.). (1999). *Tests in print V* (Vol. II). Lincoln, NE: The Buros Institute of Mental Measurements, The University of Nebraska.

PLAKE, B. S., & IMPARA, J. C. (Eds.). (2001). *The fourteenth mental measurements yearbook.* Lincoln, NE: The Buros Institute of Mental Measurements, The University of Nebraska.

SAMPSON, J. P., & LUMSDEN, J. A. (2000). Ethical issues in the design and use of Internet-based career assessment. *Journal of Career Assessment, 8,* 21–35.

SCOTT, G. S. (1998). *Making ethical choices, resolving ethical dilemmas.* St. Paul, MN: Paragon.

WELFEL, E. (2002). *Ethics in counseling and psychotherapy: Standards, research, and emerging issues.* Pacific Grove, CA: Brooks/Cole.

BIBLIOGRAPHY

AMERICAN COUNSELING ASSO-CIATION. (1995). *Code of Ethics and Standards of Practice.* Alexandria, VA: Author.

AMERICAN COUNSELING ASSO-CIATION. (1999). *Ethical standards for Internet online counseling.* Alexandria, VA: Author.

ANDERSON, B. S. (1996). *The counselor and the law* (4th ed.). Alexandria, VA: American Counseling Association.

ARTHUR, G. L., & SWANSON, C. D. (1993). *Confidentiality and privileged communication.* In T. P. Renley (Ed.), *ACA legal series* (Vol. 6). Alexandria, VA: American Counseling Association.

BETAN, E. J., & STANTON, A. L. (1999). Fostering ethical willingness: Integrating emotional and contextual awareness with rational analysis. *Professional Psychology: Research and Practice, 30,* 295–301.

BIAGIO, M., DUFFY, R., & STAFFELBACH, D. F. (1998). Obstacles to addressing professional misconduct. *Clinical Psychology Review, 18,* 273–285.

BROWN, S. P., & ESPINA, M. R. (2000). Report of the ACA Ethics Committee. *Journal of Counseling and Development, 78,* 237–241.

CHAUVIN, J. C., & REMLEY, T. P., Jr. (1996). Responding to allegations of unethical conduct. *Journal of Counseling and Development, 74,* 563–568.

Codes of ethics for the helping professions. (2003). Pacific Grove, CA: Brooks/Cole.

COHEN, E. D., & COHEN, G. S. (1999). *The virtuous therapist: Ethical practice of counseling and psychotherapy.* Pacific Grove, CA: Brooks/Cole.

COPPER, C. C., & GOTTLIEB, M. C. (2000). Ethical issues with managed care: Challenges facing counseling psychology. *The Counseling Psychologist, 28,* 179–236.

COREY, G. (2001). *Student manual for Corey's theory and practice of counseling and psychotherapy* (6th ed.). Belmont, CA: Brooks/Cole.

COTTONE, R. R., & CLAUS, R. E. (2000). Ethical decision making models: A review of the literature. *Journal of Counseling and Development, 78,* 275–283.

FALVEY, J. E. (2002). *Managing clinical supervision: Ethical practice and legal risk management.* Pacific Grove, CA: Brooks/Cole.

GOLDEN, L., & SCHMIDT, S. J. (1998). Unethical practice as perceived by mental health professionals: The next generation. *Counseling and Values, 42,* 166–170.

GRESSARD, C. F., & KEEL, L. (1998). An introduction to the ACA Code of Ethics and Standards of Practice. *Counseling Today, 40,* 16–23.

HOPKINS, W. E. (1997). *Ethical dimensions of diversity.* Thousand Oaks, CA: Sage.

HOPKINS, W. E., & ANDERSON, B. (1990). *The counselor and the law* (3rd ed.). Alexandria, VA: American Counseling Association.

LAFROMBOISE, T. D., FOSTER, S., & JAMES, A. (1996). Ethics in multicultural counseling. In P. B. Pederson, J. G. Draguns, W. J. Lonner, & J. E. Trimble (Eds.), *Counseling across cultures* (4th ed., pp. 47–72). Thousand Oaks, CA: Sage.

LINDSAY, G., & CLARKSON, P. (1999). Ethical dilemmas of psychotherapists. *Psychologist, 12,* 182–183.

MURPHY, L. L., IMPARA, J. C., & PLAKE, B. S. (Eds.). (1999). *Tests in print V* (Vol. II). Lincoln, NE: The Buros Institute of Mental Measurements, The University of Nebraska.

NATIONAL BOARD FOR CERTI-FIED COUNSELORS ETHICAL CODE. (1998). *National Board for Certified Counselors News Notes, 14.*

PEDERSEN, P. (1995). Cross cultural ethical guidelines. In J. B. Ponterotto, J. M. Casas, L. A. Suzuki, & C. M. Alexander (Eds.), *Handbook of multicultural counseling* (pp. 34–50). Thousand Oaks, CA: Sage.

PEDERSEN, P. (1997). The cultural context of the American Counseling Association Code of Ethics. *Journal of Counseling and Development, 76,* 23–28.

PLAKE, B. S., & IMPARA, J. C. (Eds.). (2001). *The fourteenth mental measurements yearbook.* Lincoln, NE: The Buros Institute of Mental Measurements, The University of Nebraska.

POPE, K. S., & VASQUEZ, M. J. T. (1998). *Ethics in psychotherapy and counseling* (2nd ed.). San Francisco: Jossey-Bass.

SAMPSON, J. P., Jr., KOLODINSKY, R. W., & GREENO, P. B. (1997). Counseling on the information highway: Future possibilities and potential problems. *Journal of Counseling and Development, 75,* 203–212.

WELFEL, E. R., & HANNIGAN-FARLEY, P. (1996). Ethics education in counseling: A survey of faculty and student views. *ICA Quarterly, 140,* 24–33.

11

Finishing Up

We anticipate that it is with great pride and a sense of accomplishment that you now approach the end of your placement. We hope that your journey in clinical growth has been a fruitful one and that you are now ready to enter the "real world" of therapy.

However, several items need attention before you exit this preprofessional experience. We have compiled a checklist of activities pertaining to clients, services, clients' continuity of care, and programmatic issues that will help you finish up the process.

CLIENT TRANSFER

- Determine when you will notify your clients of the impending completion of your internship.
- Determine when to stop accepting new clients.
- Address clients' feelings and provide opportunities to say good-bye.
- Address your own feelings about ending these treatment relationships.
- Arrange for a new counselor to be assigned.
- Schedule a counseling session for your clients to be introduced to their new counselor, when possible.

- Decide with your clients the stage of your counseling relationship conducive for the transfer to occur, whenever possible.

- Maintain precise documentation permitting the new counselor and client to know exactly where to begin services (treatment plan, transfer summary, progress notes).

- Notify the social service agencies that provided case management/ advocacy of your impending departure and give them the name of your replacement.

- Clarify the current status of issues you have been addressing with various outside agencies.

- Verify that all letters, reports, and records that are requested or needed are sent.

- Have the client sign new releases of information so exchange of information continues.

TERMINATING CLIENTS

- Attempt to achieve the goals of treatment plans for clients to be terminated.

- Determine what to do if treatment goals were not (or will not be) achieved by termination.

- Notify clients of services available to them if the need arises (such as support groups, crisis telephone help lines, and emergency counseling services).

- Adhere to the detailed provisions of after-care services (regarding medication, review appointments, support groups, bibliotherapy, and so on).

- Again, maintain precise documentation. We cannot overemphasize this point.

PROGRAM COVERAGE

- Arrange for coverage for groups.

- Remove yourself from schedules of intake assessments, emergency services, didactic programs, and research activities.

- Curtail all client and program services within two weeks of departure from the site.

- Offer your telephone number and address if the agency needs to contact you.

You should schedule an exit interview with your site supervisor and your university supervisor to review your performance and to complete the final evaluation. You will discuss your areas of competence (therapeutic orientation, theories, techniques), the clinical population with which you were most effective, and the service areas (individual, group, intake, emergency, case management) in which you demonstrated the most skill. General impressions and recommendations should be discussed at this time. Your input on your performance should be considered, and you should have sufficient opportunities to ask questions and get any needed clarification. Your supervisors should have been providing feedback on your performance during the entire placement; thus, the final evaluation should not be a time for surprises. Your comments regarding your site supervisor's role and his or her relationship with you can be discussed. Ideally, your comments on the internship experience will be welcomed and will add to your supervisor's effectiveness.

Your site supervisor and university supervisor will probably endorse your successful completion of the internship by writing an evaluation form (see Appendix M for a Sample Evaluation of Intern Form). At the conclusion of this meeting, take the opportunity to complete any necessary licensure or certification forms.

Finally, your supervisors may recommend ways to present your experiences clearly and concisely in your resume. As a practitioner, your site supervisor understands what would increase your marketability. He or she may share information about employment opportunities and career trends. The site supervisor may suggest that you network with mental health professionals regarding job opportunities and strategies. This would also be a good time to request a professional reference.

Recognize that you have an obligation to yourself to examine your own feelings regarding this termination (refer to Appendix N for a Sample Intern Evaluation of Site/Supervisor Form). After all, has this internship not been a developmental process for you? Certainly your clients have moved through a therapeutic process, but you yourself have also made an educational journey consisting of several discrete steps, including starting, acclimating (or adjusting), working, assessing, integrating, and finishing. In many ways, this process parallels the termination process for clients in therapy:

- Starting: You bring a fund of knowledge, enthusiasm, and eagerness to learn, an openness to experience our profession in the real world.

- Adjusting: You adapt to a new environment and individuals with the goal of understanding how you fit into the overall dynamics of the agency.

- Working: You are now able to use your knowledge and skills. You aspire to establish credibility as a counselor.

- Assessing: You evaluate the quality of your performance and test your understanding of the intricacies of the counseling field. In addition, you solicit feedback regarding your efforts.

- Integrating: You are able to implement the feedback you receive, coupled with practical knowledge gained, to become a more effective counselor.

- Finishing: You feel prepared to transfer the formal and practical knowledge into the counseling field.

We hope that this book has been helpful throughout your internship. We have tried to provide the practical information you needed to function in your role as counselor-in-training. We also hope that during your internship you were able to integrate your theoretical training with the practical expectations and demands of our profession.

We like our clients to take a neatly wrapped package away with them from therapy, a sort of gift to themselves, one that they can feel good about. This package contains much hard work, personal growth, independence, a sense of accomplishment and survival, and, in Tillich's (2000) words, "the courage to be." Likewise, we want you to take a neatly wrapped package away with you as you complete this field experience. Listed here are the contents of your internship gift to yourself. (Please permit us to wax philosophical.)

- May you have the sense that you have participated in a meaningful professional and educational experience that has touched you personally.

- Emerson is supposed to have written that one criterion for determining success is the knowledge that "by your having lived someone has breathed easier." May you come away with a commitment to assist others in their independence from you.

- May you develop a sense of competency: competency in skills, interactions, theoretical knowledge base, and asking for information and help when you do not know the answers.

- May you be kind and patient with yourself during this process. At the same time, may you be as anxious as you need to be to keep yourself motivated. Remember that you can't care effectively for others if you can't first take care of yourself.

- May you continue to learn, grow, and change and adapt to new and uncertain circumstances in both your personal life and your professional life, which are, after all, inextricably intertwined.

- May you leave a positive mark on others, as we hope we have done with you.

We wish you much success and happiness in our profession.

REFERENCE

TILLICH, P. (2000). *The courage to be* (2nd ed.). New Haven, CT: Yale University Press.

APPENDIX A

Sample Résumé

CORY A. JONES
2300 Main Street
Cleveland, OH 44107
(216) 555-1111
E-mail: Cory.Jones@companyname.com

Objective	An internship in the human services and counseling area
Education	B.A., Hiram College, Hiram, OH: June 2002 Major: Psychology (Departmental Honors) Cumulative GPA: 3.5
	M.A. in Community Counseling, John Carroll University, Cleveland, OH: Expected May 2006
Work Experience	*Neighboring Mental Health Center,* Mentor, OH: 2001–present Case Manager
	Cleveland Memorial Hospital, Cleveland, OH: 2000–2001 Case Aide
Honors and Activities	Dean's list, Hiram College: 1998–2001 Departmental Honors, Hiram College: 2001 Tutoring Program, Cleveland, OH: 1999 Psychology Club, Hiram College: 1998–2001 Copresenter, American Counseling Association: 2003
References	Available upon request

APPENDIX B

Sample Thank-You Note

Cory A. Jones
2300 Main Street
Cleveland, OH 44107
(216) 555-1111
E-mail:
Cory.Jones@companyname.com

Daniel Noday, PhD
Cleveland Counseling Agency
22000 Euclid Avenue
Cleveland, OH 44120

Dear Dr. Noday:

I appreciate your taking time to meet with me yesterday. I enjoyed having the opportunity to speak with you and to learn about your agency. I am interested in a field placement with the Cleveland Counseling Agency beginning in September 2004. I will call you in about a week to discuss this possibility.

Thank you in advance for your consideration.

Sincerely,

Cory Jones

APPENDIX C

Sample Internship Announcement

INTERNSHIP/PRACTICUM PLACEMENT
FOR GRADUATE STUDENTS IN COUNSELING

Available for the Fall Semester
Cleveland Mental Health Center

Opportunities for

- Mental health counseling with a wide range of clients
- Participating in current aspects of the agency's outpatient and community education programs
- Developing new services for clients

Training supervision toward Ohio licensure in counseling available.

Contact: Marc Noday, M.D.
Clinical Director
(216) 555-2200
Clevelandmhc.org

APPENDIX D

NBCC Coursework Requirements

NBCC COURSEWORK AREA DESCRIPTIONS

To meet the educational requirements for the NCC credential, you must hold an advanced degree with a major study in counseling. You must also have completed 48 semester or 72 quarter hours of graduate-level counseling coursework at a regionally accredited institution of higher education, with at least one course in each of the areas listed below. A separate course must be counted for each content area. Each course must have been completed for at least two semester hours or three quarter hours of graduate credit. Field experience must have been completed for three semester hours or five quarter hours of graduate credit. You may be required to submit a program description and/or course descriptions photocopied from your graduate catalog to determine the acceptability of your coursework. If these are also vague, syllabi may be requested. Coursework must demonstrate a counseling focus.

Human Growth and Development. Includes studies that provide an understanding of issues and needs of individuals at all developmental levels.

Social and Cultural Foundations. Includes studies that provide an understanding of issues and trends in a multicultural and diverse society.

Helping Relationships. Includes studies that provide an understanding of counseling theories and techniques.

Group Work. Includes studies that provide an understanding of group development, group dynamics, group counseling theories and techniques, and other group work approaches.

Career and Lifestyle Development. Includes studies that provide an understanding of the client's career development and related life factors.

Source: National Board for Certified Counselors. Coursework Requirements (2002). Reprinted by permission of NBCC. Please note: Guidelines were current upon publication, but applicants should contact NBCC to assure they are applying under current standards.

Appraisal. Includes studies that provide an understanding of individual and group approaches to assessment and evaluation.

Research and Program Evaluation. Includes studies that provide an understanding of types of research methods.

Professional Orientation. Includes studies that provide an understanding of all aspects of counseling functions, including history, roles, organizational structures, ethics, standards, and credentialing in the counseling progression. (Please note: Psychology, social work, human sciences, and marriage and family therapy are different professions than counseling.)

Field Experience. Refers to supervised counseling experience in an appropriate work setting that provided at least two academic terms of graduate credit. Each field experience must have been earned for at least three semester hours or five quarter hours of graduate credit. The field experience must be completed through a regionally accredited college or university. Applicants who earned only one academic term of field experience may substitute one additional year of supervised work as a counselor (1,500 extra hours of counseling experience and 50 extra hours of face-to-face supervision) beyond the required two years of post-master's counseling supervised experience required of other applicants. Both the experience and supervision must span three years for applicants choosing this option.

APPENDIX E

Technological Competencies for Community Counseling Students

ACES Technology Interest Network

At the completion of a counselor education program, students should:

1. Be able to use productivity software to develop Web pages, group presentations, letters, and reports.

2. Be able to use such audiovisual equipment as video recorders, audio recorders, projection equipment, video conferencing equipment, and playback units.

3. Be able to use computerized statistical packages.

4. Be able to use computerized testing, diagnostic, and career decision-making programs with clients.

5. Be able to use e-mail.

6. Be able to help clients search for various types of counseling-related information via the Internet, including information about careers, employment opportunities, education/training opportunities, financial assistance/scholarships, treatment procedures, and social/personal information.

7. Be able to subscribe, participate in, and sign off counseling listservs.

8. Be able to access and use counseling-related CD-ROM databases.

9. Be knowledgeable of the legal and ethical codes that relate to counseling services via the Internet.

10. Be knowledgeable of the strengths and weaknesses of counseling services provided via the Internet.

11. Be able to use the Iinternet for finding and using continuing education opportunities in counseling.

12. Be able to evaluate the quality of Internet information.

From http://www.acesonline.net/competencies.htm, Association for Counselor Education and Supervision, ACES Technical Competencies for Counselor Education Students: Recommended Guidelines for Program Development: ACES Technology Interest Network (1999). Copyright 1999 ACES.

APPENDIX F

Alphabetical List of Commonly Used Psychotropic Medications

Trade Name (Generic Name)	Class	Disorder or Target Symptoms	Side Effects	Other Facts
Abilify (*Aripiprazole*)	Antipsychotic	Schizophrenia	Agitation, headache, insomnia.	Newest antipsychotic developed and causes fewest side effects. Unique mechanism of action: dopamine system stabilizer.
Adderall (same as Concerta, Metadate, and Ritalin) (*Methylphenidate hydrochloride*)	Psychostimulant	ADD/ADHD	Appetite suppression, dizziness, dry mouth, GI upset, headache, hypertension, impotence, insomnia, irritability, tremor, palpitations, tachycardia, weight loss.	Associated with psychological dependence.
Ambien (*Zolpidem*)	Sedative/Hypnotic	Insomnia	Agitation, amnesia, dizziness, drowsiness, headache, nausea, nightmares.	Short-acting. May cause rebound effect. Some potential for abuse.
Anafranil (*Clomipramine*)	Antidepressant (Tricyclic)	OCD; major depression	Blurred vision, constipation, dizziness, drowsiness, dry mouth, EKG abnormalities, GI upset, increased appetite, irregular heartbeat, postural hypotension, sedation, tachycardia, tremor, urinary retention, weakness, weight gain.	
Aricept (*Donepezil hydrochloride*)	Cognitive Aid (Anticholinestrase Inhibitor)	Alzheimer's disease	Anorexia, diarrhea, insomnia, nausea, vomiting, fatigue, muscle cramps.	Slows progress of disease in some patients, but does not cure Alzheimer's disease.
Asendin (*Amoxapine*)	Antidepressant (Tricyclic)	Major depression	Blurred vision, constipation, dizziness, drowsiness, dry mouth, EKG abnormalities, GI upset, increased appetite, irregular heartbeat, postural hypotension, sedation, tachycardia, tremor, urinary retention, weakness, weight gain.	

Trade Name (Generic Name)	Class	Disorder or Target Symptoms	Side Effects	Other Facts
Ativan (Lorazepam)	Antianxiety (Benzodiazepine)	Anxiety	Dizziness, drowsiness, fatigue, headache, muscle cramps, nausea, vomiting, weakness.	Some potential for drug tolerance and dependence.
Benadryl (Diphenhydramine)	Antihistamine	Insomnia; treatment of side effects of antipsychotics	Dizziness, dry mouth, drowsiness, GI upset, itching, sedation, skin rash, sweating, thickening of respiratory secretions, urinary retention, wheezing.	Contraindicated in individuals with asthma, peptic ulcers, prostate problems, urinary problems.
Buspar (Buspirone)	Antianxiety (Azaspirone)	Anxiety	Agitation, dizziness, drowsiness, headaches, GI upset, insomnia.	Three-week lag before effects are felt. Ineffective for panic disorder or very severe anxiety. Adverse interaction with MAOIs. No risk for addiction.
Catapres (Clonidine)	Antihypertensive	Anxiety; aggressive behavior and ADHD in children	Constipation, dizziness, dry mouth, fatigue, sedation.	Primary use is treatment for hypertension. Abrupt withdrawal may cause agitation, anxiety, headache, hypertension.
Celexa (Citalopram)	Antidepressant (SSRI)	Major depression	Agitation, agranulocytosis, dry mouth, hypotension, insomnia, low white cell count, sexual dysfunction, somnolence, weight gain.	Dangerous drug interactions with MAOIs.
Clozaril (Clozapine)	Antipsychotic	Schizophrenia; psychotic features of mood disorder	Anxiety, constipation, convulsions, depression, dizziness, drowsiness, dry mouth, excess salivation, GI upset, headache, shortness of breath, skin rash, tachycardia, weight gain.	Contraindicated in patients with cardio-vascular, eye, kidney, liver, prostate, or urinary problems. Potentially fatal blood disease may occur. Not associated with tardive dyskinesia.
Concerta (same as Adderall, Metadate, and Ritalin) (Methylphenidate hydrochloride)	Psychostimulant	ADD/ADHD	Appetite suppression, depression, dizziness, growth suppression, GI upset, insomnia, weight loss.	Once-a-day dose. Associated with psychological dependence.

(continued on next page)

Trade Name (Generic Name)	Class	Disorder or Target Symptoms	Side Effects	Other Facts
Cylert (*Pemoline*)	Psychostimulant	ADD/ADHD	Appetite suppression, growth suppression, insomnia, weight loss.	Risk of liver damage.
Cymbalta (*Duloxetine*)	Antidepressant (Atypical)	Major depression	Dry mouth, nausea, somnolence.	Does not cause weight changes. Does not raise blood pressure. Sexual dysfunction (delayed orgasm) in males only.
Dalmane (*Flurazepam*)	Antianxiety (Benzodiazepine)	Anxiety; insomnia	Dizziness, drowsiness, fatigue, headache, muscle cramps, nausea, vomiting, weakness.	Adverse interaction with alcohol. Some risk for addiction.
Depakote (*Valproic acid*)	Mood Stabilizer (Anticonvulsant)	Bipolar; migraine	Appetite decrease, blood disorders, GI upset, itching, menstrual changes, tardive dyskinesia.	Potentially fatal liver damage may occur. Local irritation of mouth may occur if tablets are chewed before swallowing.
Desyrel (*Trazodone*)	Antidepressant (Atypical)	Major depression; anxiety; insomnia	Dizziness, dry mouth, postural hypotension, sedation.	Risk of sustained erection (priapism) in men, requiring emergency medical intervention.
Effexor (*Venlafaxine*)	Antidepressant (Atypical)	Major depression; ADD/ADHD	Appetite changes, constipation, dizziness, dry mouth, GI upset, headache, insomnia, sedation, somnolence, sweating, tachycardia.	Effective treatment for melancholic depression; monitor blood pressure at doses ≥150 mg/day.
Elavil (*Amitriptyline*)	Antidepressant (Tricyclic)	Major depression; chronic pain	Blurred vision, constipation, dizziness, drowsiness, dry mouth, EKG abnormalities, GI upset, increased appetite, irregular heartbeat, postural hypotension, sedation, tachycardia, tremor, urinary retention, weakness, weight gain.	
Eskalith (**same as Lithobid**) (*Lithium carbonate*)	Mood Stabilizer	Bipolar	Acne, ataxia, confusion, drowsiness, dry mouth, GI upset, hair loss, hallucinations, memory problems, thyroid problems, tremor, urinary problems, weight gain.	Narrow margin of safety; toxic or lethal if too high; monitor blood levels, thyroid function, and kidney function.

Trade Name (Generic Name)	Class	Disorder or Target Symptoms	Side Effects	Other Facts
Exelon (*Revastigmine*)	Cognitive Aid (Anticholinestrase Inhibitor)	Alzheimer's disease	Abdominal pain, dizziness, headache, nausea, somnolence, tremor, weakness, weight loss, vomiting.	Slows progress of disease in some patients, but does not cure Alzheimer's disease.
Geodon (*Ziprasidone*)	Antipsychotic	Schizophrenia; psychotic features of mood disorder	Abnormal muscle movements, constipation, cough, diarrhea, dizziness, fatigue, restlessness, rhinitis, skin rash, tremor.	Only antipsychotic that is associated with little or no weight gain. Controls both positive and negative symptoms of psychosis. Risk of tardive dyskinesia not yet known.
Halcion (*Triazelam*)	Sedative/Hypnotic (Benzodiazepine)	Insomnia	Dizziness, drowsiness, fatigue, headache, muscle cramps, nausea, vomiting, weakness.	
Halcion (*Triazelam*)	Antianxiety (Benzodiazepine)	Insomnia	Dizziness, drowsiness, fatigue, headache, muscle cramps, nausea, vomiting, weakness.	
Haldol (*Haloperidol*)	Antipsychotic	Schizophrenia; psychotic features of mood disorder; Tourette's	Agitation, akathisia, akinesia, anxiety, depression, dizziness, dry mouth, excess salivation, fatigue, insomnia, nasal congestion, rigidity, sedation, tremor, weight changes.	Associated with tardive dyskinesia.
Imovane (*Zopiclone*)	Sedative/Hypnotic	Insomnia	Agitation, amnesia, dizziness, drowsiness, headache, nausea, nightmares.	Short-acting. May cause rebound effect. Some potential for abuse.
Inderal (*Propanolol*)	Antihypertensive	Anxiety	Depression, fatigue, hypotension, sedation, tachycardia.	Contraindicated for individuals with asthma or diabetes.
Kava Kava (*Piper mythystocum*)	Herbal	Anxiety; insomnia	Somnolence; long-term use (more than three months) may result in skin rash, blood abnormalities, facial swelling, muscle weakness.	Increases side effects of antianxiety drugs, antidepressants, sleep aids, pain relievers, muscle relaxants. May result in coma when used with alprazolam.

(continued on next page)

Trade Name (Generic Name)	Class	Disorder or Target Symptoms	Side Effects	Other Facts
Klonopin (*Clonazepan*)	Antianxiety (Benzodiazepine)	Bipolar; aggression; panic; anxiety; insomnia; Tourette's; restless legs syndrome; aggressive behavior in children	Dizziness, drowsiness, fatigue, headache, muscle cramps, nausea, vomiting, weakness.	
Lamictal (*Lamatrogine*)	Mood Stabilizer (Anticonvulsant)	Bipolar	Ataxia, blurred vision, cough, dizziness, nausea and vomiting, rhinitis, skin rash, somnolence, sore throat.	Risk of accidental injury due to poor coordination and dizziness. Must discontinue if skin rash develops.
Lexapro (*Escitalopram*)	Antidepressant (SSRI)	Major depression	Agitation, blurred vision, dizziness, drowsiness, nausea, sexual dysfunction, weight changes.	Dangerous drug interactions with MAOIs. Lexapro is the active isomer of Celexa.
Librium (*Chlordiazepoxide*)	Antianxiety (Benzodiazepine)	Anxiety	Dizziness, drowsiness, fatigue, headache, muscle cramps, nausea, vomiting, weakness.	Slow onset. Some risk for addiction.
Lithobid (same as Eskalith) (*Lithium carbonate*)	Mood Stabilizer	Bipolar	Acne, ataxia, confusion, drowsiness, dry mouth, GI upset, hair loss, hallucinations, memory problems, thyroid problems, tremor, urinary problems, weight gain.	Narrow margin of safety; toxic or lethal if too high; monitor blood levels; regulate liquid intake and output.
Loxitane (*Loxapine*)	Antipsychotic	Schizophrenia; psychotic features of mood disorder	Akathisia, blurred vision, dizziness, dry mouth, fatigue, hair loss, headache, hypotension, nasal congestion, pseudo-Parkinsonism, tachycardia, tardive dyskinesia, tremor, urinary retention.	Contraindicated in patients with epilepsy. Associated with tardive dyskinesia.
Ludiomil (*Maprotaline*)	Antidepressant (Tetracyclic)	Major depression	Blurred vision, dry mouth, constipation, sedation, skin rash, weight gain.	
Luvox (*Fluvoxamine*)	Antidepressant (SSRI)	OCD	Anxiety, appetite decrease, dry mouth, GI upset, insomnia, sedation, sexual dysfunction, tremor.	Dangerous drug interactions may occur with Xanax, Hismanal, Seldane, Halcion. Abrupt withdrawal may result in headache, dizziness, nausea.

Trade Name (Generic Name)	Class	Disorder or Target Symptoms	Side Effects	Other Facts
Mellaril (Thioridazine)	Antipsychotic	Schizophrenia; psychotic features of mood disorder	Blurred vision, cardiac arrhythmias, constipation, dizziness, drowsiness, dry mouth, hypotension, nasal congestion, sexual dysfunction, skin rash, swelling, tardive dyskinesia, urinary retention, weight gain.	Adverse interactions with anesthetics, barbiturates, narcotics. Avoid insecticides. Associated with tardive dyskinesia.
Metadate (same as Adderall, Concerta, and Ritalin) (Methylphenidate hydrochloride)	Psychostimulant	ADD/ADHD	Appetite decrease, headache, insomnia.	Once-a-day dose. Associated with psychological dependence.
Nardil (Phenelzine)	Antidepressant (MAOI)	Major depression	Blurred vision, dizziness, dry mouth, edema, excitement, GI upset, hypertension, hypertensive crisis, hypomania, insomnia, postural hypotension, sedation, sexual dysfunction, tachycardia, tremor, urinary hesitancy or retention, weight gain.	Dangerous interactions with SSRIs and many other Rx and OTC drugs and foods. Abrupt discontinuation may cause agitation, anxiety, nightmares, psychosis. Hypertensive crisis is acute medical emergency.
Navane (Thiothixene)	Antipsychotic	Schizophrenia; psychotic features of mood disorder	Akathisia, dizziness, drowsiness, dry mouth, hypotension, nasal congestion, pseudo-Parkinsonism, skin rash, tardive dyskinesia, urinary retention, weight gain.	Adverse interactions with anesthetics, barbiturates, narcotics. Associated with tardive dyskinesia.
Neurontin (Gabapentin)	Mood Stabilizer (Anticonvulsant)	Bipolar; anxiety; neuropathic pain	Ataxia, dizziness, double vision, fatigue, rhinitis, somnolence.	Take at bedtime to decrease side effects.

(continued on next page)

Trade Name (Generic Name)	Class	Disorder or Target Symptoms	Side Effects	Other Facts
Norpramin (Desipramine)	Antidepressant (Tricyclic)	Major depression; ADHD	Blurred vision, constipation, dizziness, drowsiness, dry mouth, EKG abnormalities, GI upset, increased appetite, irregular heartbeat, postural hypotension, sedation, tachycardia, tremor, urinary retention, weakness, weight gain.	
Pamelor (Nortriptyline)	Antidepressant (Tricyclic)	Major depression; ADHD	Blurred vision, constipation, dizziness, drowsiness, dry mouth, EKG abnormalities, GI upset, increased appetite, irregular heartbeat, postural hypotension, sedation, tachycardia, tremor, urinary retention, weakness, weight gain.	
Parnate (Tranylcypromine)	Antidepressant (MAOI)	Major depression	Dizziness, dry mouth, constipation, excitement, GI upset, headaches, hypertension, hypertensive crisis, hypomania, insomnia, postural hypotension, sedation, tachycardia.	Dangerous interactions with SSRIs and many other Rx and OTC drugs and foods. Abrupt discontinuation may cause agitation, anxiety, confusion, depression, delirium, headaches, insomnia, muscle weakness, nightmares, psychosis. Hypertensive crisis is acute medical emergency.
Paxil (Paroxetine)	Antidepressant (SSRI)	Major depression; OCD; panic	Constipation, dizziness, dry mouth, GI upset, headaches, sedation, sexual dysfunction, sweating.	Flu-like syndrome may occur upon discontinuation unless dose is tapered gradually.
Phenobarbital (Phenobarbital)	Antianxiety (Barbiturate)	Anxiety	Agitation, anxiety, ataxia, dizziness, confusion, GI upset, hallucinations, headaches, impaired thinking, insomnia, nightmares, somnolence, swelling of facial features.	Hypersensitivity may produce severe, potentially fatal skin condition. Liver damage may occur. High risk for addiction.

Trade Name (Generic Name)	Class	Disorder or Target Symptoms	Side Effects	Other Facts
Prolixin (*Fluphenazine*)	Antipsychotic (Phenothiazine)	Schizophrenia; psychotic features of mood disorder	Agitation, blurred vision, drowsiness, dry mouth, excess salivation, GI upset, insomnia, sexual dysfunction, tardive dyskinesia, weight gain.	Adverse interactions with alcohol, epinephrine, hypnotics. Contraindicated in patients with liver disease. Associated with tardive dyskinesia.
Provigil (*Modafinil*)	Psychostimulant	Narcolepsy	Anxiety, appetite suppression, headache, infection, insomnia, nausea.	Used off-label to treat ADD/ADHD and Alzheimer's disease.
Prozac (**same as Sarafem**) (*Fluoxetine*)	Antidepressant (SSRI)	Major depression; OCD; panic, premenstrual dysphoric disorder	Anxiety, GI upset, headaches, insomnia, sedation, sexual dysfunction, tremor, weight loss.	Longest half-life of SSRIs. Dangerous drug interactions with MAOIs; wait five weeks after stopping Prozac before taking MAOIs. No research evidence to support increased suicidality, despite media reports (Stahl, 1996). Once-a-week dose available.
Remeron (*Mirtazapine*)	Antidepressant (SSRI)	Major depression	Appetite increase with weight gain, dizziness, elevated cholesterol and triglycerides, sedation, sexual dysfunction, somnolence.	Dangerous drug interactions with MAOIs; wait 14 days after stopping Remeron before taking MAOIs.
Reminyl (*Galantamine HBr*)	Cognitive Aid (Anticholinestrase Inhibitor)	Alzheimer's disease	Appetite decrease with weight loss, diarrhea, dizziness, GI upset, muscle weakness, sweating, vomiting.	Slows progress of disease in some patients, but does not cure Alzheimer's disease.
Restoril (*Tenazepam*)	Antianxiety (Benzodiazepine)	Anxiety; insomnia	Dizziness, drowsiness, fatigue, headache, muscle cramps, nausea, vomiting, weakness.	Some risk for addiction.
Risperdal (*Risperidone*)	Antipsychotic	Schizophrenia; psychotic features of mood disorder; borderline personality	Constipation, dizziness, drowsiness, tardive dyskinesia, weight gain.	Controls both positive and negative symptoms. Avoid alcohol. Risk of tardive dyskinesia not yet known.

(continued on next page)

Trade Name (Generic Name)	Class	Disorder or Target Symptoms	Side Effects	Other Facts
Ritalin (same as Adderall, Concerta, and Metadate) (Methylphenidate hydrochloride)	Psychostimulant	ADD/ADHD	Appetite suppression, depression, dizziness, growth suppression, GI upset, insomnia, weight loss.	Associated with psychological dependence and drug tolerance.
St. John's Wort (Hypericum)	Herbal	Mild to moderate major depression; anxiety	Restlessness.	Lowers effectiveness of birth control pills and drugs used to treat HIV (protease inhibitors and nonnucleoside reverse transcriptase inhibitors). May increase side effects of SSRIs and MAOIs. Increases risk of sunburn.
Sarafem (same as Prozac) (Fluoxetine)	Antidepressant (SSRI)	Premenstrual dysphoric disorder	Anxiety, constipation, GI upset, headaches, insomnia, sedation, sexual dysfunction, tremor, weight loss.	Same drug as Prozac. Longest half-life of SSRIs. Dangerous drug interactions with MAOIs; wait five weeks after stopping Sarafem before taking MAOIs.
Serax (Oxazepam)	Antianxiety (Benzodiazepine)	Anxiety	Dizziness, drowsiness, fatigue, headache, muscle cramps, nausea, vomiting, weakness.	Slow onset. Some risk for addiction.
Seroquel (Quetiapine)	Antipsychotic	Schizophrenia; psychotic features of mood disorder; borderline personality	Dizziness, drowsiness, dry mouth, headache, tardive dyskinesia.	Controls both positive and negative symptoms. Avoid alcohol. Risk of tardive dyskinesia not yet known.
Serzone (Nefazodone)	Antidepressant (Atypical)	Major depression	Dizziness, dry mouth, headache, sedation, somnolence.	Does not cause anxiety or insomnia. Low risk of mania-induction in bipolar. Some risk of liver failure.

Trade Name (Generic Name)	Class	Disorder or Target Symptoms	Side Effects	Other Facts
Sinequan (Doxepin)	Antidepressant (Tricyclic)	Major depression	Blurred vision, constipation, dizziness, drowsiness, dry mouth, EKG abnormalities, GI upset, increased appetite, irregular heartbeat, postural hypotension, sedation, tachycardia, tremor, urinary retention, weakness, weight gain.	
Sonata (Zaleplon)	Sedative/Hypnotic	Insomnia	Agitation, amnesia, dizziness, drowsiness, headache, nausea, nightmares.	Short-acting. May cause rebound effect.
Stelazine (Trifluoperazine)	Antipsychotic (Phenothiazine)	Schizophrenia; psychotic features of mood disorder	Akathisia, appetite decrease, cardiac arrest, constipation, dizziness, drowsiness, hair loss, headache, muscle weakness, pseudo-Parkinsonism, skin rash, tardive dyskinesia, tremor.	Adverse interactions with alcohol, anesthetics, antianxiety agents, barbiturates, narcotics. Associated with tardive dyskinesia.
Strattera (Atomoxetine)	Noradrenaline Reuptake Inhibitor	ADD/ADHD	Dry mouth, GI upset, insomnia, weight loss.	Newest FDA-approved medication for ADD/ADHD.
Surmontil (Trimipramine)	Antidepressant (Tricyclic)	Major depression	Blurred vision, constipation, dizziness, drowsiness, dry mouth, EKG abnormalities, GI upset, increased appetite, irregular heartbeat, postural hypotension, sedation, tachycardia, tremor, urinary retention, weakness, weight gain.	Toxic in overdose.
Tegretol (Carbamazepine)	Mood Stabilizer (Anticonvulsant)	Bipolar	Ataxia, blood disorders, blurred vision, constipation, dizziness, dry mouth, drowsiness.	May precipitate mania. Adverse interaction with MAOIs. Risk for rare but lethal skin and blood conditions. Monitor with blood tests.
Tenex (Guanfacine)	Antihypertensive	Aggressive behavior and ADHD in children	Constipation, dizziness, drowsiness, dry mouth.	

(continued on next page)

Trade Name (Generic Name)	Class	Disorder or Target Symptoms	Side Effects	Other Facts
Tenormin (*Atenolol*)	Antihypertensive	Anxiety	Depression, dry eyes, headache, sexual dysfunction, sedation, skin rash, sore throat.	Contraindicated for individuals with kidney disease. Abrupt withdrawal may cause cardiac problems.
Thorazine (*Chlorpromazine*)	Antipsychotic (Phenothiazine)	Schizophrenia; psychotic features of mood disorder	Akinesia, blurred vision, constipation, depression, dry mouth, drowsiness, EKG abnormalities, menstrual changes, muscle weakness, pseudo-Parkinsonism, skin pigmentation, tachycardia, tardive dyskinesia, weight gain.	Adverse interactions with alcohol, anesthetics, antianxiety agents, anticoagulants, anticonvulsants, barbiturates, epinephrine, narcotics. Associated with tardive dyskinesia.
Tofranil (*Imipramine*)	Antidepressant (Tricyclic)	Major depression; bedwetting in children	Blurred vision, constipation, dizziness, drowsiness, dry mouth, EKG abnormalities, GI upset, increased appetite, irregular heartbeat, postural hypotension, sedation, tachycardia, tremor, urinary retention, weakness, weight gain.	Increased risk of falls in elderly.
Topamax (*Topiramate*)	Mood Stabilizer (Anticonvulsant)	Bipolar	Anxiety, breast pain in women, constipation, dizziness, drowsiness, fatigue, GI upset, nausea, vision problems.	May cause change in sense of taste.
Tranxene (*Clorazepate*)	Antianxiety (Benzodiazepine)	Anxiety	Dizziness, headache, skin rash.	Moderate risk for addiction.
Trilafon (*Perphenazine*)	Antipsychotic	Schizophrenia; psychotic features of mood disorder	Akathisia, blurred vision, constipation, dizziness, dry mouth, nasal congestion, pseudo-Parkinsonism, somnolence, tardive dyskinesia, tremor, weight gain.	Contraindicated in patients with depression, kidney, or liver problems. Adverse interactions with alcohol, antihistamines, analgesics, barbiturates, opiates. Associated with tardive dyskinesia.

Trade Name (Generic Name)	Class	Disorder or Target Symptoms	Side Effects	Other Facts
Trileptal (Oxcarbazepine)	Mood Stabilizer (Anticonvulsant)	Bipolar; chronic pain	Dizziness, double vision, drowsiness, hyponatremia.	Does not require monitoring of blood levels or liver function. May reduce effectiveness of birth control pills.
Valerian Root (Valeriana officinalis)	Herbal	Anxiety; insomnia; premenstrual dysphoric disorder	Somnolence, dizziness. Long-term use may cause excitability, headaches, insomnia, restlessness, tachycardia.	Increases side effects of muscle relaxants, sleep aids, antianxiety drugs, pain relievers, antidepressants.
Valium (Diazepam)	Antianxiety (Benzodiazepine)	Anxiety	Dizziness, drowsiness, fatigue, headache, muscle cramps, nausea, vomiting, weakness.	Rapid action. High risk for addiction.
Vistaril (Hydroxyzine)	Antihistamine	Anxiety	Drowsiness, dry mouth, sedation, tremor.	Adverse interactions with alcohol, analgesics, barbiturates, narcotics.
Vivactil (Protriptyline)	Antidepressant (Tricyclic)	Major depression	Blurred vision, constipation, dizziness, drowsiness, dry mouth, EKG abnormalities, GI upset, increased appetite, irregular heartbeat, postural hypotension, sedation, tachycardia, tremor, urinary retention, weakness, weight gain.	
Wellbutrin (same as Zyban) (Bupropion)	Antidepressant (Atypical)	Major depression; ADHD	Agitation, anxiety, appetite/weight changes, constipation, insomnia, sweating.	Low incidence of sexual dysfunction. Increased risk of seizure at high doses (>400 mg), with initial rapid increase in dosage, or in clients having bulimia or seizure disorders.
Xanax (Alprazolam)	Antianxiety (Benzodiazepine)	Anxiety; panic	Dizziness, drowsiness, fatigue, headache, muscle cramps, nausea, vomiting, weakness.	Some risk for addiction.

(continued on next page)

Trade Name (Generic Name)	Class	Disorder or Target Symptoms	Side Effects	Other Facts
Zoloft (*Sertraline*)	Antidepressant (SSRI)	Major depression; OCD; panic; PTSD; social phobia	Dizziness, dry mouth, GI upset, insomnia, sexual dysfunction, somnolence, sweating, tremor.	Dangerous drug interactions with MAOIs; wait 14 days after stopping Zoloft before taking MAOIs. Abrupt withdrawal may result in headache, dizziness, nausea.
Zyban (same as Wellbutrin) (*Buproprion*)	Antidepressant (Atypical)	Smoking cessation	Agitation, anxiety, appetite and weight changes, constipation, insomnia, sweating.	Increased risk of seizure at high doses (>400 mg), with initial rapid increase in dosage, or in clients having bulimia or seizure disorders.
Zyprexa (*Olanzapine*)	Antipsychotic	Schizophrenia; psychotic features of mood disorder; borderline personality	Akathisia, constipation, dizziness, postural hypotension, somnolence, weight gain.	Controls both positive and negative symptoms. Risk of tardive dyskinesia not yet known.

APPENDIX G

State and National Credentialing Boards, Professional Organizations, and Honorary Societies

STATE CREDENTIALING BOARDS

ALABAMA
Board of Examiners in Counseling
P.O. Box 550397
Birmingham, AL 35255
(205) 933-8100
(205) 933-6700 (fax)

ALASKA
Counselor Licensure Board
Board of Professional Counselors
Department of Community and
Economic Development
Division of Occupational Licensing
P.O. Box 110806
Juneau, AK 99811-0806
(907) 465-2551
(907) 465-2974 (fax)
www.dced.state.ak.us/occ/ppco.htm

ARIZONA
Counselor Credentialing Committee
of the Board of Behavioral Examiners
1400 W. Washington Street, Suite 350
Phoenix, AZ 85007
(602) 542-1882
(602) 542-1830 (fax)
www.bbhe.state.az.us/

ARKANSAS
Board of Examiners in Counseling
Southern Arkansas University
P.O. Box 1396
Magnolia, AR 71753
(501) 235-4314
(501) 234-1842 (fax)
akthomas@saummag.edu

Source: American Counseling Association.

CALIFORNIA
Board of Behavioral
Science Examiners
400 R Street, Suite 3150
Sacramento, CA 95814-6240
(916) 445-4933
(916) 323-0707 (fax)

COLORADO
Board of Licensed Professional
Counselor Examiners
1560 Broadway, Suite 1340
Denver, CO 80202
(303) 894-7766
(303) 894-7790 (fax)
amos.martinez@state.co.us

CONNECTICUT
Connecticut Department of
Public Health
410 Capitol Avenue, MS #12 APP
P.O. Box 340308
Hartford, CT 06134
(860) 509-7579

DELAWARE
Board of Professional Counselors of
Mental Health
861 Silver Lake Blvd., Cannon
Building, #203
Dover, DE 19904
(302) 739-4522
(302) 739-2711 (fax)

DISTRICT OF COLUMBIA
DC Board of Professional Counselors
614 H Street NW, Room 108
Washington, DC 20001
(202) 727-5365
(202) 727-4087 (fax)

FLORIDA
Board of Clinical Social Workers
Marriage & Family Therapy & Mental
Health Counseling
Department of Health, Medical
Quality Assurance
2020 Capital Circle SE, Bin #C08
Tallahassee, FL 32399-3250
(850) 487-1129
(850) 921-2569 (fax)
Note: Fax documents are not
accepted.
www.doh.state.fl.us/mqa/491/
soc_home.htm

GEORGIA
Composite Board of Professional
Counselors, Social Workers, and
Marriage & Family Therapists
166 Pryor Street SW
Atlanta, GA 30303
(404) 656-3933
(404) 651-9532 (fax)

HAWAII
No information available at this time.

IDAHO
Idaho State Counselor
Licensure Board
Bureau of Occupational Licenses
1109 Main Street, Suite 220
Boise, ID 83702-5642
(208) 334-3233
(208) 334-3945 (fax)
www2.state.id.us/ibol/cou.htm

ILLINOIS
Professional Counselor &
Disciplinary Board
320 W. Washington Street
Springfield, IL 62786
(217) 785-0872
(217) 782-7645 (fax)

INDIANA
Social Worker, Marriage & Family
Therapist, & Mental Health
Counselor Board
Health Professions Bureau
402 W. Washington Street, Room 041
Indianapolis, IN 46204
(317) 232-2960
(317) 233-4236 (fax)
www.IN.gov/hpb/boards/mhcb

IOWA
Iowa Behavioral Science Board
321 E. 12th Street, Lucas Bldg.,
4th Floor
Des Moines, IA 50319
(515) 281-6352
(515) 281-3121 (fax)

KANSAS
Behavioral Sciences Regulatory Board
712 S. Kansas Avenue
Topeka, KS 66603-3817
(913) 296-3240
(913) 296-3112 (fax)
www.ksbsrb.org

KENTUCKY
Board of Professional Counselors
Occupations & Professions
Perry Hall Annex
P.O. Box 456
Frankfort, KY 40602
(502) 564-3296, ext. 226
(502) 564-4818 (fax)
www.kyca.org

LOUISIANA
Licensed Professional Counselors
Board of Examiners
8631 Summa Avenue, Suite A
Baton Rouge, LA 70809
(504) 765-2515
(504) 765-2514 (fax)

MAINE
Board of Counseling Professionals
State House, Station #35
Augusta, ME 04333
(207) 624-8603
(207) 624-8637 (fax)
DIANESTAPLES@STATE.ME.US

MARYLAND
Board of Examiners of
Professional Counselors
Metro Executive Center, 3rd Floor
4201 Patterson Avenue
Baltimore, MD 21215
(410) 764-4732
(410) 764-5987 (fax)

MASSACHUSETTS
Board of Allied Mental Health &
Human Services Professionals
239 Causeway Street
Boston, MA 02114
(617) 727-3080
(617) 727-2366 (fax)
www.state.ma.us/reg/boards/mh

MICHIGAN
Michigan Board of Counseling
P.O. Box 30670
Lansing, MI 48909
(517) 335-0918
(517) 373-3596 (fax)

MINNESOTA
No information available at this time.

MISSISSIPPI
Mississippi Board of Examiners
for LPCs
319 S. Main Street
Yazoo City, MS 39194
(888) 860-7001

MISSOURI
Missouri Committee for
Professional Counselors
P.O. Box 1335
Jefferson City, MO 65102
(573) 751-0018
(573) 751-4176 (fax)

MONTANA
Board of Social Work Examiners &
Professional Counselors
301 S. Park, 4th Floor
P.O. Box 200513
Helena, MT 59620-0513
(406) 841-2369
(406) 841-2305 (fax)
www.discoveringmontana.com/dli/
bsd/license/bsd_boards/swp_board/
board_page.htm

NEBRASKA
Board of Examiners in
Professional Counseling
301 Centennial Mall South
P.O. Box 95007
Lincoln, NE 68509
(402) 471-2117
(402) 471-0380 (fax)

NEVADA
No information available at this time.

NEW HAMPSHIRE
New Hampshire Board of Examiners
in Psych. & Mental Health Practice
105 Pleasant Street, #457
Concord, NH 03301
(603) 271-6762

NEW JERSEY
New Jersey Professional Counselor
Examiner's Committee
Division of Consumer Affairs
P.O. Box 45033
Newark, NJ 07101
(973) 504-6415
(973) 648-3536 (fax)

NEW MEXICO
Counselor Therapy & Practice Board
P.O. Box 25101
Santa Fe, NM 87504
(505) 827-7160
(505) 827-7085 (fax)

NEW YORK
No information available at this time.

NORTH CAROLINA
North Carolina Board of Licensed
Professional Counselors
P.O. Box 21005
Raleigh, NC 27619-1005
(919) 787-1980
(919) 571-8672 (fax)

NORTH DAKOTA
North Dakota Board of
Counselor Examiners
2112 10th Avenue SE
Mandan, ND 58554
(701) 224-8234
(701) 667-5969 (fax)

OHIO
Counselor & Social Worker Board
77 S. High Street, 16th Floor
Columbus, OH 43266-0340
(614) 752-5161
(614) 644-8112 (fax)

OKLAHOMA
Licensed Professional Counselors,
Licensed Marital & Family Therapists
1000 NE 10th Street
Oklahoma City, OK 73117-1299
(405) 271-6030
(405) 271-1918 (fax)
www.health.state.ok.us/program/lpc/

OREGON
Board of Licensed Professional
Counselors and Therapists
3218 Pringle Road SE, #160
Salem, OR 97302-6312
(503) 378-5499

PENNSYLVANIA
State Board of Social Workers,
Marriage & Family Therapists, &
Professional Counselors
P.O. Box 2649
Harrisburg, PA 17105
(717) 783-1389
(717) 787-7769 (fax)

RHODE ISLAND
Board of Mental Health Counselors &
Marriage & Family Therapists
3 Capitol Hill, Cannon Bldg.,
Room 104
Providence, RI 02908-5097
(401) 222-2827 ext. 106
(401) 222-1272 (fax)

SOUTH CAROLINA
South Carolina Department of LLR
Division of POL
Board of Examiners for LPC, AC
and MFT
P.O. Box 11329
Columbia, SC 29211
(803) 734-4243
(803) 734-4284 (fax)

SOUTH DAKOTA
South Dakota Board of
Counselor Examiners
P.O. Box 1822
1116 S. Minnesota Avenue
Sioux Falls, SD 57101-1822
(605) 331-2927
(605) 331-2043 (fax)
www.state.sd.us/dcr/counselor

TENNESSEE
State Board of Professional Counselors
& MFTs
1st Floor, Cordell Hull Bldg.
426 5th Avenue North
Nashville, TN 37247-1010
(888) 310-4650
(615) 532-5164 (fax)

TEXAS
Board of Examiners of
Professional Counselors
1100 W. 49th Street
Austin, TX 78756-3183
(512) 834-6658
(512) 834-6677 (fax)

UTAH
Division of Occupational and
Professional Licensing
160 E. 300 South, 4th Floor
P.O. Box 45802
Salt Lake City, UT 84111
(801) 530-6789
(801) 530-6511 (fax)

VERMONT
Board of Allied Mental
Health Practitioners
Office of Professional Regulations
Secretary of State's Office
Red Stone Bldg.
26 Terrace Street, Drawer 09
Montpelier, VT 05609-1106
(802) 828-2390
(802) 828-2496 (fax)

VIRGINIA
Board of Counseling
Department of Health Professions
6606 W. Broad Street, 4th Floor
Richmond, VA 23230-1717
(804) 662-9912
(804) 662-9943 (fax)

WASHINGTON
Counselor Programs
Department of Health
P.O. Box 47869
Olympia, WA 98504-7869
(360) 664-9098
(360) 586-7774 (fax)

WEST VIRGINIA
Board of Examiners in Counseling
Ona, WV 25545
(800) 520-3852
(304) 767-3061
(304) 767-3062 (fax)

WISCONSIN
Examining Board of Social Workers,
MFTs & Professional Counselors
P.O. Box 8935
Madison, WI 53708-8935
(608) 267-7212
(608) 267-0644 (fax)

WYOMING
Occupational Licensing Director
Mental Health Professions
Licensing Board
2301 Central Avenue
Cheyenne, WY 82002
(307) 777-7788
(307) 777-3508 (fax)

NATIONAL COUNSELOR CERTIFICATION

National Board for Certified Counselors
3 Terrace Way, Suite D
Greensboro, NC 27403
(910) 547-0607
(910) 547-0017 (fax)
nbcc@nbcc.org
www.nbcc.org

PROFESSIONAL ASSOCIATION

American Counseling Association
5999 Stevenson Avenue
Alexandria, VA 22304-3300
(800) 347-6647
(703) 823-9800
(703) 823-0252 (fax)
www.counseling.org

HONORARY SOCIETY

Chi Sigma Iota Counseling Academic and Professional Honor Society
International School of Education
250 Ferguson Bldg.
University of North Carolina at Greensboro
Greensboro, NC 27412-5001
(910) 334-4035
www.csi-net.org

ACCREDITATION AGENCY

Council for Accreditation of Counseling and Related Educational Programs
5999 Stevenson Avenue
Alexandria, VA 22304
(703) 823-9800 ext. 301
(703) 823-0252 (fax)
(703) 370-1943 (TTD)

APPENDIX H

Commonly Used Abbreviations

@	about or at
AA	Alcoholics Anonymous
a.c.	before meals
ACSW	Academy of Certified Social Workers
ad. lib.	as much as desired
adm.	admitted
adol.	adolescent
AMA	against medical advice
amt.	amount
a.o.	anyone
appt.	appointment
as.	of each
ASA	against staff advice
ASAP	as soon as possible
ASR	at staff request
B/4	before
b/c	because
b.f.	boyfriend
b.i.d.	twice a day
b.i.n.	twice a night
bl.	black
B.M.	bowel movement
B.P. or B/P	blood pressure

bro.	brother
-c or w/	with
c	canceled
Ca	calcium
CA	cancer
CAI	Career Assessment Inventory
cap.	capsule
CAPS	Career Ability Placement Survey Test
C.B.C.	complete blood count
cc	cubic centimeter
CCDC	Certified Clinical Drug Counselor
CCMHC	Certified Clinical Mental Health Counselor
cl or ct.	client
CNS	central nervous system
c/o	complains or complained of
C.P.T.	Common Procedural Terminology
C/R	cancel/reschedule
Cx	cancel
D or dos.	dose
D & A	drug and alcohol
DAT	Differential Aptitude Test
d.c.	discontinue
detox	detoxification
dis.	discontinued
DNA	did not appear
dr. or z	dram
Dr.	doctor
DSM	Diagnostic and Statistical Manual
D.T.'s	delirium tremens
Dx	diagnosis
E	evolved
ECT	electroconvulsive therapy
EEG	electroencephalogram
e.g.	for example
EKG or ECG	electrocardiogram
e.o.	everyone
ER, EW, or EMR	emergency room or emergency ward
exam.	examination
fa.	father
fl. or fld.	fluid
F.R.	fully resolved
ft.	feet
GA	Gamblers Anonymous
g.f.	girlfriend
GI	gastrointestinal
gm.	gram

gma.	grandmother
gmpa. or gpa.	grandfather
gr.	grain
GRP	group
gtt.	drop
H2O	water
H & P	history and physical exam
h.s.	hour of sleep or bedtime
ht.	height
Hx	history
ICDM 10	*International Classification of Diseases Manual* (Vol. 10)
i.e.	that is
Iden.	identification
I.M.	intramuscular
in.	individual
info.	information
ISB	Incomplete Sentences Blank
IV	intravenous
JCAHO	Joint Commission on Accreditation of Healthcare Organizations
K	potassium
L.	left
Lge.	large
liq.	liquids
LPC	Licensed Professional Counselor
LPCC	Licensed Professional Clinical Counselor
LPN	Licensed Practical Nurse
mcg.	microgram
m Eq.	milliequivalent
mg.	milligram
Mg	magnesium
Mg S04	magnesium sulfate
min.	minute
ml.	milliliter
mm.	millimeter
MMPI	Minnesota Multiphasic Personality Inventory
mo.	mother
mod.	moderate
mot.	motivation
MS	mental status
Mtg.	meeting
Na	sodium
N.A.	Narcotics Anonymous
NA	not applicable
n/ach	not achieved

NCC	National Certified Counselor
neg.	negative
n.o.	no one
no. or #	number
NPO	nothing by mouth
ns	no show
obs.	observation
O.B.S.	organic brain syndrome
o.d.	everyday
O.D.	overdose
oint.	ointment
o.m.	every morning
OTO	one time only
oz. or z	ounce
p	present
p.c.	after meals
pil.	pill
p.o.	by mouth
P.O.	probation officer
P.R.	partially resolved
prn	whenever necessary, as needed
pt.	patient
q.	every
q.2h.	every two hours
q.4h.	every four hours
q.d.	every day
q.h.	every hour
q.i.d.	four times a day
q.o.	every other
q.o.d.	every other day
q.s.	quantity sufficient
R. or R	right
re:	regarding
rec.	received
R.N.	Registered Nurse
R/O	rule out
Rx	prescription, medication
s	without
s.c.	subcutaneously
schiz.	schizophrenia
sib.	sibling
sig.	label it
sis.	sister
s.o.	someone
S.O.	significant other
sol.	solution

s.o.s.	if necessary
ss	one half
stat.	immediately
sx	symptom
tab.	tablet
tbsp.	tablespoon
T/C	telephone call
temp.	temperature
t.i.d.	three times a day
t.i.n.	three times a night
T.O.	telephone order
T.P.R.	temperature, pulse, respiration
tr.	tincture
tsp.	teaspoon
Tx	treatment
U.	unit
U.R.	utilization review
UR	unresolved
V.O.	verbal order
w/ (or –c)	with
w/i	within
wky.	weekly
w/o	without
wt.	weight
x	times
∴	therefore
↑	increase
↓	decrease
♂	male
♀	female

APPENDIX I

Intern Field Log/Journal

Intern's Name _____

Internship Site _____

Date	Direct Hours	Indirect Hours	Indiv. Super. Hours	Group Super. Hours	Description of Internship Activities	My Reactions

Current Totals:

____   ____   ____   ____

Previous Totals:

____   ____   ____   ____

Accumulative Totals:

____   ____   ____   ____

_____
Supervisor's Signature

Thanks,
Cynthia Osborn, PhD
Kent State University

APPENDIX J

Sample Intake Form

Client Name_____ Counselor Name_____

Date_____ Length of Interview_____

I. Client and Concern Description: *Often the formula denoting age, race, gender, marital status in adults, followed by the words "complaining of . . ." or "reporting . . .";for example, a 54-year-old Caucasian male, married, complaining of . . . or reporting . . .*

II. Psychosocial History: *Note here relevant historical data on the client, exploring such areas as childhood and adolescence, current family and family of origin history (including children), legal history, medical history, religious history, employment history, educational history, military history, history of mental health contacts, etc.*

III. Mental Status: Client appeared alert (y,n) and oriented (y,n).

Cognitions: *The counselor explores current functioning in the cognitive arena, including memory, intellectual level, concrete versus abstract thinking abilities, evidence of hallucinations, delusions, etc.*

Affect: *Here the counselor notes current type, level, and intensity of emotions, including depression, anxiety, mania, etc.*

Behaviors: *In this area of mental status, the counselor takes special notice of any behavioral anomalies, including tics, psychomotor agitation or retardation, pressured speech, unusual gestures, etc.*

Risk of Harm to Self or Others: *This area of current functioning is of special import and thus is listed separately. The counselor needs to determine level of lethality and any and all ideation, plans, and means. We strongly suggest that the intern become familiar with agency and ethical policies in this area.*

IV. Diagnostic Impression (DSM-IV)

Axis I	Clinical Syndromes
	Other Conditions That May Be a Focus of Clinical Attention
Axis II	Personality Disorders
	Mental Retardation
Axis III	General Medical Conditions
Axis IV	Psychosocial and Environmental Problems
Axis V	Global Assessment of Functioning

V. Treatment Recommendations: *As with the diagnostic impression, treatment recommendations are conclusional data resulting from the process of the interview. And as with the diagnostic impression, these recommendations are tentative and subject to modification as needed. They may include recommendations for psychological testing; psychiatric referral for medication assessment; referral to another professional; a specific type of therapy, such as marital, family, individual, or group; a specific modality, such as hypnosis, biofeedback, or cognitive/behavioral; hospitalization; or even no treatment. A treatment plan (see Appendix K) usually follows. We suggest that the client be as involved in the process as much as possible.*

VI. Additional Remarks:

APPENDIX K

Sample Treatment Plan

Client Name _____

Counselor Name _____

Date _____

Problem/ Concern	Goal	Treatment Methods	Expected Date of Achievement	Results	Follow-up
1. depressive symptoms	learn to i.d. auto negative thoughts	cognitive-behavioral strategies	01/2006		
2. list others here					
3.					
4.					
5.					

(This treatment plan follows a management-by-objective format. Note that it is specific, time-limited, and concise. Further, it allows for client input in the therapy process. Notice, too, that the goal is usually the inverse of the problem statement. We recommend that the client sign the form along with the counselor; thus it becomes a social contract. In the case of minors, we have them and their parents sign, allowing for some concrete investment in therapy by both minor and parent. Finally, we have discovered that managed care companies like such a tangible document.)

_____ _____ _____ _____
Client   Intern   Supervisor   Parent (as necessary)

APPENDIX L

Therapy Session Case Note Components

1. Type of treatment (e.g., Ind. Tx).

2. Length of session (e.g., 1 hr.).

3. Mini Mental Status Examination (standard formula: e.g., "Ct. appeared alert, oriented")
   Hint: Note any aberrations in cognition, affect, and behavior.

4. What was reported? This is information reported by the client.
   Hint: Use ct.'s quotes as appropriate

5. What was discussed? This is information obtained during the session and is reported like "minutes of the meeting." Again, ct.'s quotes may be helpful.

6. Give homework if possible.
   Hint: Consider small, discrete tasks to help ensure successes.
   Hint: Homework should be completed by next session.

7. Date/time of next session.

8. Clinician's signature, degree, license, credential (e.g., Sally Smith, MA, LPC, NCC).

Nota bene:

- One clinician we know of reviews notes with the ct. as appropriate in an effort to establish a collaborative (horizontal versus vertical) relationship. However, some agency policies do not allow for this.
- If client is suicidal, but not imminently so, you may want to consider inserting a statement (in essence, a behavioral contract) to this effect:

  "I will not try to kill myself without first calling the emergency # and/or clinician, and I will go to a hospital emergency room if I cannot reach either party."

_____   _____   _____   _____
Client signs   Counselor signs   Supervisor signs   Date

Give a copy of this section to the client.

Insert direct quotes from the client to support this clinical stance.

Always follow policies and procedures of the agency, provided they are legal and ethical.

APPENDIX M

Sample Evaluation
of Intern Form

Intern's name _____

Date of this evaluation_____   Name of agency _____

Agency address _____

Agency phone/E-mail address _____

Univ. supervisor _____   Agency supervisor _____

Title _____   Title _____

Professional degree _____   Professional degree _____

Licensed as_____ No. _____   Licensed as_____ No. _____

Certified as_____ No. _____   Certified as_____ No. _____

Internship dates __/__/__ to __/__/__   Intern paid? _____

Total number of placement hours _____

Total number of supervisory hours _____

Suggested Competencies for Interns
4 = Outstanding   3 = Good   2 = Fair   1 = Poor   NA = Not applicable

   I. Communication skills
      a. Verbal skills _____
      b. Writing skills _____
      c. Knowledge of nomenclature _____
      Comments:

II. Interviewing
    a. Structure of interview ____
    b. Attending behaviors ____
    c. Active listening skills ____
    d. Professional attitude ____
    e. Interviewing techniques ____
    f. Mental status evaluation ____
    g. Psychosocial history ____
    h. Observation ____
    i. Use of questions ____
    j. Reflection ____
    k. Empathy ____
    l. Respect for differences ____
    Comments:

III. Diagnosis
    a. Knowledge of assessment instruments ____
    b. Knowledge of current DSM ____
    c. Use of records ____
    d. Ability to formulate a preliminary diagnosis ____
    Comments:

IV. Treatment
    a. Ability to draw up a treatment plan ____
    b. Ability to perform individual counseling ____
    c. Ability to perform marital counseling ____
    d. Ability to perform conjoint counseling ____
    e. Ability to perform family counseling ____
    f. Ability to perform group counseling ____
    g. Crisis intervention skills ____
    h. Ability to perform brief models of counseling ____
    i. Consultation skills ____
    j. Ability to deal with various populations/diversity issues ____
    k. Ability to make progress notes ____
    Comments:

V. Case management
    a. Knowledge of agency programs and professional staff roles ____
    b. Knowledge of community resources ____
    c. Discharge planning ____
    d. Follow-up ____
    e. Record keeping of client management ____
    f. Collaboration with other agencies, which also serve clients ____
    Comments:

VI. Agency operations and administration
    a. Knowledge of agency mission, structure, and organization ____
    b. Awareness of roles of administrative staff ____
    c. Knowledge of agency goals ____
    d. Understanding of agency care standards, including managed care ____
    e. Use of technology ____
    Comments:

VII. Professional orientation
    a. Knowledge of counselor ethical codes ____
    b. Knowledge of agency professional policies ____
    c. Ability of intern to seek and accept supervision ____
    Comments:

Please write a brief summary statement of the intern as a future counselor:

Intern _____

Agency Supervisor _____

University Supervisor _____

Date _____

APPENDIX N

Sample Intern Evaluation of Site/Supervisor Form

Intern's name_____

Date of this evaluation_____ Name of agency_____

Agency supervisor _____ Title & license _____

Internship dates __/__/__ to __/__/__ Total # hours _____

THE SITE
4 = Outstanding   3 = Good   2 = Fair   1 = Poor   NA = Not Applicable

   I. Overall agency operations and administration _____

  II. Introduction to agency mission and structure _____

 III. Awareness of roles of administrative staff _____

 IV. Knowledge of agency goals _____

  V. Understanding of agency care standards _____

 VI. Policies on duty to risk management (including duty to warn) _____

VII. Policies on confidentiality of records _____

Comments:

THE SUPERVISOR

4 = Outstanding   3 = Good   2 = Fair   1 = Poor   NA = Not Applicable

Noting the competency areas suggested for my internship (refer to Appendix M, Evaluation of Intern Form, and Chapter 2), I might rate my supervisor's assessment and/or training of me in these areas as:

   I. Communication skills _____

  II. Interviewing _____

 III. Diagnosis (including knowledge of DSM–IV) _____

 IV. Treatment (including treatment planning and termination) _____

  V. Case management _____

 VI. Agency operations and administration _____

VII. Professional orientation _____

VIII. Knowledge and application of professional ethics _____

 IX. Ability to process my professional issues with supervisor _____

Comments:

Index